Basic High School
Math Review

Basic High School
Math Review

Jim Elander

To order additional copies of this book, contact:
Xlibris LLC
1-888-795-4274
www.Xlibris.com
Orders@Xlibris.com
131417

Contents

High School Math Review with Decision Making Skills
(It is assumed you have had Algebra and Geometry.)

Session 1   Some Old and Some New...7

Session 2   Review 1...16

Session 3   Implications and Applications 21

Session 4   The use and misuse of Basic Statistics 31

Session 5   Applications with Problem Solving 41

Session 6   Implications and Polls..53

Session 7   3-D Applications 68

Session 8   Applications: Some old, some new...............................76

Session 9   Algebra and Number Activities
             (Some different types of thinking!) 85

Session 10   Epilogue or The Final Word 100

Session 11   Additional Selected Activities for Math Review.............107

Index 1   Basic Definitions 119
Index 2   Postulates..125
Index 3   Essential Geometry Theorems 127
Index 4   Conversion Information 131
Index 5   Suggestions for further reading 133
Index 6   Quotes...139

Session 1

Some Old and Some New

You either make your own decisions or you become a slave to other decision makers.

Unknown

A Democracy's success depends on an educated voting public.

Unknown

Parents (Teachers) should not teach their children (students) what to think, but teach them How to think and make decisions.

Elander

Session 1. Introduction

Most of the students who will use this Review and apply to an institution which usually requires some type of admission information and regardless of your plans for the future, you will always be making decisions.

All entrance assessments contain questions related to Geometry, but many students have forgotten most of their Geometry and Algebra. So this review will contain a basic review and applications of these subjects plus provide the basis for Decision Making or Critical Thinking as related to everyday situations.

In the following set of questions you will be given a few examples of the mathematical questions and a reminder that all conclusions are based on undefined terms, defined terms, basic assumptions and previous conclusions (deductive reasoning), plus making conclusions from a few previous cases (inductive reasoning). (If you don't understand Deductive or Inductive Reasoning and how they are used in decision making, this review will aid you.)

Examples:

U.S. Constitutional Amendment XIX.

The right of citizens of the United States to vote shall not be denied or abridged by the United States or by any State on account of sex.

(The answers are the author's. Your answers may differ due to rounding.)

What does "abridged" mean? (This may be an undefined term to you.)

What is the Total number of words? # _____   28

How many would you classify as Undefined? # _____   17

How many would you classify as Defined? # ____   11

What is the percent of the total number of words that you classified as undefined?

> Hint: The number of "undefined words" (The, of, to, etc.) is what percent of the total number of words?
> > Write the question in equation form and solve.
> > > Undefined # = x% of total # $17 = x\%(28)$
> > > > $17 = x(1/100)(28)$
> > > > Where did the 1/100 come from?
> > > > Solving: $x = 60.7\%$

Questions: 1. What number (numeral?) does the symbol % represent? (What is the difference between number and numeral? We usually don't differentiate between the to terms.)
2. What number (numeral?) does XIX represent? 19
3. What year was the XIX Amendment adopted? 1919-20

From the above, we could "jump" to a general conclusion that about 60% of the Constitution consists of undefined terms. You should question that general conclusion since it is based on only one small case. (This is an example of Inductive Reasoning. What does inductive reasoning mean to you from this one example?)

An example of Deductive Reasoning is: Some States in the South passed laws that required a tax in order to vote, therefore town X required a tax in order to vote. (This is an example of Deductive Reasoning, which is drawing a conclusion from a general statement to a specific case.)

This was ruled unconstitutional since it violated the XIX Amendment.

(Decisions (good or bad) are base on the following:
  Inductive Reasoning
  Deductive Reasoning
    Direct and Indirect
  What you see?
  What you read and the quality of the material.
  What you listen to or don't listen to on the radio, or
      lectures, conversations, or even gossip, and don't question
      the qualifications of the persons providing the information.
  What you watch on TV and its validity.
  Advertisements
  Polls and their results
    (You need to know how the poll information was
    collected, when, where, and how it was collected,
    who was asked, the wording of the questions, plus
    the number interviewed.)
  Implications and their forms with interpretations
    (Converses, Inverses, and Contrapositives)

    (The above will all be explained, reviewed and
    applied later.)

An interesting example of a conclusion related to "What we see." is illustrated in the following. The following picture will illustrate this. Look at it from the right and then from the left. Which do you see, the face of an old woman, or a young woman? What if you were a witness to an accident? What would you report?

(I have no idea where following clever picture came from.)

What is seen may be reported differently by two witnesses!

A logical system is based on undefined terms, defined terms, assumptions or postulates, and theorems or laws which are justified by the first four, and of course decisions resulting from them. (A great example of a logical non-mathematical system is the Declaration of Independence by Jefferson.) (Have you read it? If not I suggest you take 15 minutes and read the first few pages.)

You will understand the full meaning of the above statement by the end of this review.

Following are two postulates for a logical Geometry system.

Postulate. 4: The shortest distance between two points is the measure of the straight line segment. (This is not always true as every taxicab driver knows.)

Example: A taxicab driver has two options to travel from A to B.)

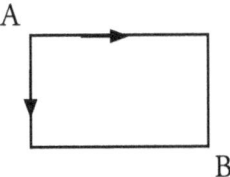

Postulate 5: The shortest distance from a point to a line is the perpendicular distance. (This also may not true in the taxi world and in earth navigation.)

Activity 1:

1. Use your ruler to draw the following triangles, with sides x, y, and z as given in centimeters.

	x	y	z
a.	3	3	5
b.	4	4	4
c.	3	4	5
d.	6	3	4
e.	6	1	5

In each case compare x + y related to z, x + z related to y, and y + z related to x.

From the comparisons, complete the following:

The sum of the measures of any two sides (segments) of a triangle is . . .

Which postulate is this based on?

Try to draw the triangle with sides of 2 in, 1 in, and 3 in.

Try to draw the triangle with sides of 2 in, 4 in, and 1 in.

If x + y > z, then is x > z - y valid?

Using the results from number 1 and given a triangle with sides 10, 18 and x, then what do you know about the length of side x?

Complete the two following inequalities where x, y and z are the sides of a plane triangle.

1. x+ y ? z   2. z ? x ? y

Hint: In a triangle, the sum of the measures of two sides is greater . . . . third side, and the third side is greater than . . . . the . . . . two other . . . sides.

Hint: From the postulate, we know that x + y > z. Can you use algebra to arrive at z > |x-y| Why the absolute value symbol?

2.　In the two figures below, how many ways can a taxi driver select to travel from A to B? (Condition: The cab must be always moving in the direction of B.)

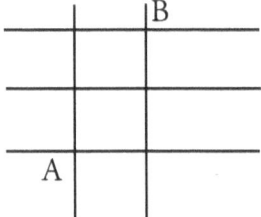 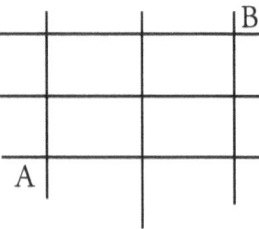

Answers: 3   6

3.　Complete the following and then write a conclusion.
    a.　9 times 1917 = ? Add the digits in the product to a one-digit number.
       Answer: The product in 'a' is 17253. Adding results is 9.
    b.　9 x 2468 = ? Adding results = ?
    c.　9 x 13579 = ? Adding results = ?
    d.　9 x 968573 = ? Adding results = ?

    e.   Write a conclusion?

    f.   Do you think 9870654321 divisible by 9?

In this session the examples illustrated that conclusions are arrived by using inductive reasoning, algebra, and the role of definition (some words may need to be defined).

4. Give an example or a case where the perpendicular distance is the shortest distance, and where it is not the shortest.

5. If $(5x + 7y - 10)/z = 25$, then which of the numbers x, y or z cannot be zero?

6. An ad is:
The professional golfers use ball T.
What does the ad want you to conclude?

7. Johnny only watches Lion News, what will he conclude?

8. If you are a U. of MT student, then you wear a Griz Cap to the game. Which of the following are valid: (If S then GC.)
    a. Joe wears a Griz Cap to the game, then he is a U of MT student.
    b. Johnny is not a U of MT student, then he does not wear a Griz Cap to the game.
    c. Joe does not wear a Griz Cap to the game, then he is not a U of MT. student.

9. Algebra
Solve the following equation for the value of x.
$$5(2x + 6) - (2x - 9) = -2$$

10. Check your answer in #9.

11. Pete, Joe, and Tom were accused of causing a problem at a school party. The principal asked the Math teacher to help solve the case as to which student is guilty. Each made a statement and only one of the statements was true. The statements are:

Pete: Joe did not do it.

Joe: Tom did it.

Tom: Joe is lying.

One of these boys is guilty, and we know only one statement is true. Indirect reasoning is to list all the possibilities and show that all cases are false or leads to a contradiction except one.

List the possibilities, record the revised statements and carefully read the results and detect the guilty student. (This is the indirect method.)

(Why is the following a theorem?)

Theorem 1: In a triangle, the sum of the measures of two sides is greater than the measure of the third side, and the third side is greater than the difference of the two other sides.

Answers:

1b. greater than the third side. d. and e. impossible to draw
   f. $a + b > c$, $a + c > b$, $b + c > a$ g. yes h. $8 < x < 28$

2. 3ways, 6 ways

3. Yes, since the sum of the digits is 45 which add to 9 and therefore the number is divisible by nine. This test is called casting out nines.

4. Taxi cab service, city A on one side of a hill to city B on the other side.

5. z can not be zero

6. The ad wants you to think that using ball T will make you a better golfer.

7. You will only hear the Lion's side of the issues.

8. a. This is the converse and is possibly invalid or valid.
   b. This the inverse and is possibly invalid or valid.
   c. This the contrapositive and is valid.

(The forms of an implication and their validities will be clarified in another session.)

9. Sloving the equation:
    $5(2x+6)-2(x-4) = -2$ (using the distributive property)
    $10x +30-2x +8 = -2$ (simplifying)
    $8x +38 = -2$ (adding a-38 to each side)
    $8x = -40$ (dividing by 8)
    $x = -5$

10. Checking
    $5(10 +6)-2(-5-4)$ should equal-2 and it does.

11. Who did it?
    Case 1. Assume Pete's statement is true and the others false.
            Pete: Joe did not do it.
            Joe: Tom did not do it.
            Tom: Tom did not do it
            Therefore: the Math teacher reported that Pete did it.
    Note: Write the other 2 cases and observe the impossibilities.

It has been said that we use indirect reasoning the most to arrive at conclusions.

A practical example is the case where the forecast is for possible rain.

Case 1: It will rain. Case 2: It doesn't rain.

Should you or shouldn't you take an umbrella.

Session 2

Review 1

Understanding evolves from work, appreciation is from applications.

Unknown

Rodin's *The Thinker (Golden Gate State, San Francisco, CA,*
(From the Elander File)

(Some Math you may have forgotten!)

Geometry Review: Short answer questions-use ruler for drawings

Suggestion: Do not check your answers until you complete the review. (It is suggested you print out the 4 indices for quick reference.)

1. Are the following valid definitions? Defend your answer.
    (A definition is valid, if the definition is true when reversed.)
    a. A restaurant is a place that serves food.
    b. Mathematics is a useful course.
    c. A postulate is an assumption.
    d. A plane triangle is a set of three non-collinear points and the line segments determined by the three points.

2. What are the three geometric terms in 1d that are classified as undefined?

3. How many points are needed to determine a geometric line?

4. The points on a line correspond to the numbers on the _____ ____ line.

5. Draw a ray and label it AB.

6. Draw a line segment and label it AC.

7. Draw a triangle and label it CDE.

8. A geometric plane is determined by ___ ___ ____ points.

9. What is the sum of the angles in a plane triangle?

10. How is the distance between points A and B determined?

11. What is a theorem?

12. What are the conditions for 2 triangles to be similar?
    a.   b.   c.

13. When are triangles congruent?

14. a. What are parallel lines?
    b. What are skew lines? (Need a dictionary?)

15. Draw three acute scalene triangles, label each ABC. (Use your ruler and protractor.)
    a. In one triangle draw the three medians.
    b. In the second triangle draw the three altitudes.
    c. In the third triangle draw the three angle bisectors.
    d. Write three conclusions.

16. The shortest distance from a point to a line is the _____ distance.

17. If A implies B and A is given, then ____.

18. Draw a rhombus that is not a square. (Need a dictionary?)

19. Draw: a. convex polygon. b. concave polygon (Need a dictionary?)

20. If the sides of a triangle are 46, 23 and x, then what do you know about the measure of x?

21. The number of square units a plane figure contains is its _____.

22. The Pythagorean Right Triangle Theorem states _____.

23. What is the sum of the interior angles in each of the following figures? (Do not use a protractor. Hint: A triangle is 180 degrees)

a.    b.    c.

24. If each letter represents a unique digit, then what are the possible digits for the following addition problem?
GO   Hints: G must > than ___ (digit) and O can't be ___?
+GO   What digit is W?
WIN

25. How many planes may four points determine?

26. The two top teams in the conference ended the season with records as shown in the following table.

Team	Wins	Losses
H	29	7
V	32	8

Team H defeated W once and Team W defeated H once. Which team would you say should receive the Cup. Why?

27. What is the perimeter of right triangle ABC, where C is the right angle?
Given: AB = 3x units and BC = 4x units

28. What is wrong with this argument?
    A yard is 36 inches.
    ¼ yard is 9 inches. (Take the square root.)
    Therefore ½ yard is 3 inches.

29. Students of school X voted that all students will wear a RED cap at
    the Saturday game. Which of the following are valid. Hint: If-Then
    form.
    a. John wore a red cap, therefore he is a student of X.
    b. John is not a student of school X, therefore he will not wear a
        red cap to the game.
    c. John did not wear a red cap to the game, therefore he is not a
        student of school X.

Answers

1. a. True, but not a definition. b. True, but not a definition. c. Yes, a
    definition. d. Valid
2. Point, line, and plane
3. Two
4. Real number
5. A   B
    →

6. A———C   7. C  D
8. Three non-collinear points   9. 180 degrees
10. Absolute value of A-B
11. A theorem is an important mathematical statement that can be
    proved.
12. a. Angles are equal (AA)
    b. The corresponding sides are proportional. (SSS)
    c. Sides proportional and included angles equal (SAS)
13. Triangles are congruent if they are similar and the ratio of sides is 1.
14. a. Parallel lines are lines in the same plane and do not intersect.
    b. Skew lines are lines not in the same plane and do not intersect.
15. Each intersect in a point 16. Perpendicular distance

17. Then B

18.   19.

20. $23 < x < 69$ units 21. Area

22. In a right triangle with sides a, b, and c, then $a^2 + b^2 = c^2$ where c is the hypotenuse.

23. a. 360 b. 720 c. 540

24. 4, 0 25. Four ABC, ABD, ACD, BCD,

26. Team H has a higher winning percentage, but there could other factors to be considered like Team W's and H's schedules, or the scores when they played each other.

27. 12x units 28. Listen to their explanations

29. "c" is valid, the contrapositive.

Suggestion: If you missed any of these problems, record the missed problem numbers, understand your errors, check the indices to refresh your memory and wait a week or two and retake the review. This review has 9 sessions and when you complete them your final test score will amaze you. **(The proof of the pudding is in the eating.)** What does that statement mean? There is nothing wrong with missing a problem once or twice, but you should learn from your errors.

Session 3

Implications and Applications

You cannot fake in mathematics, no one can be fooled.
You can either solve . . . or you cannot.

Jerry P. King(edited)
THE ART OF MATHEMATICS

Rodin's *The Thinker (Golden Gate State, San Francisco, CA,*
(From the Elander File)

Here are possibly some new forms of a mathematical statement which is very often misused in **everyday decision making**. It is the math related form that enables us to determine the validity of the conclusion. A statement in everyday situations is very often translated to the form of "If A then B" as you may recall referring to Theorems in your Geometry class. This form of a statement can be diagramed in the following manner. You can see that if in A, then in B.

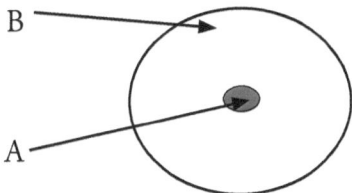

You encountered this method of writing statements as Geometry theorems and probably did not apply it to non-geometric statements, or to everyday statements and applications. If the statement is translated or converted to If-Then form many times people will make the following translations without thinking: (A simple case will illustrate these translations.) Assume the original statement converted to If-Then from is: If you have a dog, then you have a 4-legged animal. (Which is a true statement.)

Sometimes people rewrite the above statement in the form *If B THEN A*, and assume it is true. This is called the converse of the original Statement and the above would read as: *If you have a 4-legged animal, then you have a dog*. This no doubt could be false (You could have a horse.) and is many times falsely assumed to be a valid statement.

Another translation is: *If you don't have a dog, then you don't have a 4-legged animal*. This is called the Inverse of the original statement. This is also no doubt a possible false statement (You could have a horse.) and is classified as an invalid statement.

Another translation is:　If you don't have a 4-legged animal, then you don't have a dog. This is call the contrapositive and is classified as true and valid.

The above is summarized below.
    Original statement: If A, then B.　　　Assume true &valid
    Converse: If B, then A.　　　　　　　Truth and validity is questionable.
    Inverse: If not A, then Not B.　　　　Truth and validity is questionable.
    Contrapositive: If not B, then not A. True and Valid

Another simple statement to check your understanding of the above is:
    If you live Montana, then you live in the United States.
    The converse is:
        If you live in the United States, then you live in Montana.
            Is this true or valid?
    The inverse is:
        If you don't live in Montana, then you don't live in the Unites States.
            Is this true or valid?
    Write the contrapositive and classify it as to truth or validity.

The statements get much harder to classify in everyday applications and some people get angry due to false interpretations and conclusions, especially during political campaigns.

Another example:

(See diagram on page 21.)

The original statement is: If you are veteran, *then you fly the flag on the 4th of July.* Area represents the veterans and the area inside the large circle represents the set of all who fly the flag on the 4th.

Assume this statement to be true and answer the following.

Randy wrote the converse: *If you fly the flag on the 4th, then you are a veteran.*

He noticed the neighbor was flying the flag and concluded the neighbor was a veteran. You can see the conclusion may be false.

Sandy knew her neighbor was not a veteran and concluded that was why he does not fly the flag on the 4th. This is the inverse form of the original statement and is:

If not a veteran, then you don't fly the flag on the 4th.

You know this is false. Sandy made a false conclusion.

Another person twisted the statement to read: If you don't fly the flag on the 4th, then you are not a veteran. (This is called the Contrapositive and is valid assuming the original is true. (To make the problem more serious and one that actually happened results when you change the word Veteran to Patriot and check the converse, inverse, and contrapositive.)

Again the summary:

Original statement: If A then B.   Assume true &valid
Converse: If B then A.   Truth and validity is questionable
Inverse: If not A then Not B.   Truth and validity is questionable
Contrapositive: If not B then not A. True and valid depending on
                                       the classification of the original
                                       statement.

Reminder: Statements are true or false, conclusions are valid or invalid. Statements over time may change their truth classification.

(Example: No man has been to the moon was once true, but now is false.)

Another example: (Similar to one used during the early days of the Iraq war)
If you wear the flag pin, then you are an "American."
(Original)
If you are an "American," then you wear the flag pin.
(Converse)
If you didn't wear the flag pin, then you are not an American.
(Inverse)
If you are not an American, then you don't wear the flag pin.
(Contrapositive)

Many people do not understand these forms of an implication and falsely accused others, in the above case, as non—Americans or not patriotic during the early days of the Iraq War and the sad part was that some news media indirectly promoted the false interpretation.

Review 3: Applications
Directions: Use your notes, the information in the index, your calculator and other tools, but organize your work and so you can justify your answers. Answers are at the end of the activity. Hint for the solutions are listed and where to find them in one of the indices. It is suggested you print out the first 4 indices for easy referral, if you haven't already done so.

1. Equilateral triangle related questions
    a. Draw an equilateral triangle with sides measuring 2 inches using your ruler and protractor. Label the vertices A, B, C and the opposite sides a, b, and c.
    b. Calculate the perimeter.
    c. Calculate the area.
    d. Add the segments called the medians to your drawing in "a." Label them AD, BE, and CF.
    e. Label the point the medians intersection in as G.
    f. What is the number of degrees in angle AGB?
    g. What is the area of triangle AGC?

h. What is the length of segment ED?

i. Calculate the length of segment BG?

j. What is the physical property of point G? Hint: Center of____.

k. What is the measurement of angle ADB?

Hints: Index 1: Def. 3, 4, 7, 9, 16, 17, 18,

Index 3: Theorems 3, 10, 15, 20, 21, 24,

2. A city park is triangular in shape with sides measuring 900, ***, and 700, all in feet. The one measure for the side of the triangle was burred ands the mayor asked if you had an idea but it is. What is your answer?

Hint: Index 3: Theorem 1.

3. A park is in the shape of an isosceles triangle with the equal sides each measuring 150 feet and the equal angles are each 30 degrees. What is the length of the third side to nearest foot? Hint: Draw the figure.

Hints: Index 1: Def. 11,17,

Index 2: Theorems 21and 22:

4. If a square has sides measuring 10 meters, then what is the length of each diagonal?

Hint: Index 3:Theorem 21

5. The school principal needs to know the height of the school's flagpole. You are given the following facts:

Hints: Index 1: Def. 7 Index 3: Theorems 6

From a point 75 feet from the base of the pole two students observe the following;

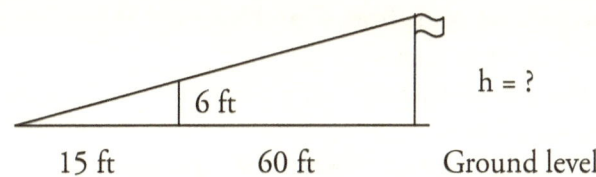

What is the height of the pole to the nearest foot?

Hint: Index 1: Def. 7

6. Given: If you are an insurance agent, then you have studied mathematics. (This is a true statement and valid)
   a. Write the converse of the given statement and indicate if it is true and valid.
   b. Write the inverse of the given statement and indicate if it is true and valid.
   c. Write the contrapositive of the given statement and indicate if it is true and valid.
   Hints: Index 1:Def. 19 Index 3: Theorem 42.

7. Induction: A person plans a garden and has 120 feet of fencing. What is the length and width of the largest rectangular area that can be fenced? In other words, the length times the width is to be a maximum.
   Suggestion: Create a table like the one below to help lead you to make your decision by inductive reasoning.
   Hint: 2L + 2W equals 120 and the area is L times W.

	Length	Width	Area
Try	5	55	275
Try	6	54	?
Try	10	?	?

   Continue until you arrive at a conclusion. Write your conclusion.

8. Study your solution to #7, then predict the maximum area if the amount of fencing were 360 feet. (Inductive reasoning—see suggestion for number 7.)

9. Draw a 30-60 degree right triangle. Label the vertices A, B, and C, where C is the vertex of the right angle and A is 60 degrees. The hypotenuse is 10 units. Hint: Index 3: Theorems 6, 20.
   a. Draw the altitude from C and label it CD.
   b. How many triangles are formed?
   c. If the triangles are similar, then map the triangles and indicate the size of each angle.
   d. If the measure of the hypotenuse is 10, then what is the measure of the sides in the three triangles? (d,e,and f)

10. Why may a four-legged table wobble?
    Hint: Index 2: Post. 6.
    (Everyday thinking type problems.)

11. Using the forms of implications to arrive at conclusions.
    Remember: If A → B (Valid)
              If B → A. Converse (Not necessarily valid)
              If not A → not B. Inverse (Not necessarily valid)
              If not B→ not A Contrapositive (Valid)
    Now apply the above to the following and indicate which you think
    are T, F or Valid or invalid.

Let A be Successful and B represent not Lazy. Name each of the
following given the original statement is: If you are Successful, then
you are not Lazy. Hint: Index 1: Def. 30.
    a. If you are not successful, then you are lazy.
    b. If you are lazy, then you are not successful.
    c. If you are not lazy, then you are successful.

12. Let F be "fun to be with" and H be "happy friends"
    If you are fun to be with, then you have happy friends.
    (F → H) Assume this is valid.
    a. Translate: If H then F. Is it always valid? Why?
    b. Translate: If not F, then not H. Is it always valid? Why?
    c. Translate: If Not H, then not F. Is it always valid? Why?

13. Given: 1. All cowboys wear handmade boots.(C→B or: If you
    are a cowboy, then you wear handmade boots.
              2. Tex is wearing handmade boots.
    Which of the following is a valid possible conclusion and why?
    a. Tex is a cowboy.   b. Tex may be a cowboy.
    c. Tex is not a cowboy.
    Hint: Write the converse, the inverse, and the contrapositive of
    statement 1 in the given and compare with the above.

The objective for problems 11-13 is to point out the validity of the
forms of an implication. These types are used in everyday thinking,
mostly without the understanding of each. Recall: Statements are
true or false and conclusions are valid or invalid.

Answers to Review 3

1   a. Drawing

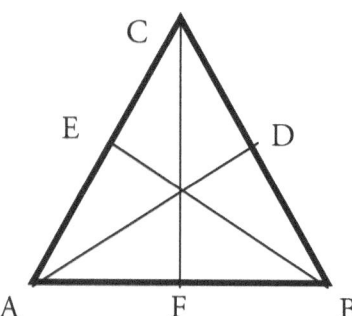

Hint: It may help if you label side AC as b, side BC as a, and c is side AB

b. P = 6 inches c. √3 or 1.732 sq in. d. Drawing e. Drawing f. 120 degrees

g. 1√3/3 or .58 sq. in. h. 1 in. i. (2/3)√3 or 1.15 j. Center of gravity

k. 90 degrees

2. 1600 > *** > 200.

3. 260 ft.
   Hint: look for a 30-60 right triangle (Th 7.2 See Index 3)

4. 10√2 or 14.4 ft.

5. 39 ft. to the nearest foot

6. a. Converse: If you studied mathematics, then you are a life insurance salesman. False and not valid.
   b. Inverse: If you aren't a life insurance salesman, then you did not study mathematics. False and invalid.
   c. Contrapositive: If you did not study mathematics, then you are not life insurance salesman. True and Valid.

7. 30 ft. by 30 ft. and area is 900 sq. ft.

8. 90 by 90 ft. and Area is 8100 sq. ft.

9. Draw right triangle ACD and altitude CD.
   Angle measurement in degrees are □A = 60 □B = 30 □C = 90, then
   a. □ACD = 30,
   b. □BCD = 60?,
   c. □ADC and □BDC = 90

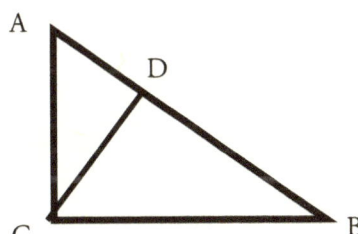

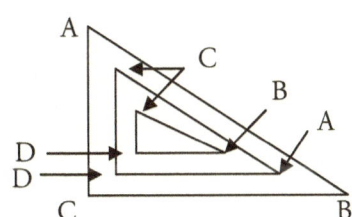

Mapping

Also is given the Measurements AB = 10.00 AC = 5.00 BD = 7.50
   d. BC = 8.60 e. CD = 4.33 f. AD = 2.50

10. Four points can determine four planes since three non collinear points will determine 1 plane. The four planes are:
   ABC, ABD, BCD, ACD if the 4 points are A, B, C, D.

11. a. is the converse—false and not valid.
    b. the contrapositive—true and valid.
    c. the inverse—false and not valid

12. a. not valid (Converse If F →H) but probably true.
    b. not valid (Inverse—not F →then not H) not valid but probably true
    c. Valid (Contrapositive—not H →not F) Valid and probably true.
       (Remember validity of a conclusion does not mean the statement is true or false.)

13. b. Is Valid and Tex may be a cowboy (converse may possibily be valid.)

Diagram also explains it, since other people can wear good boots other than cowboys.

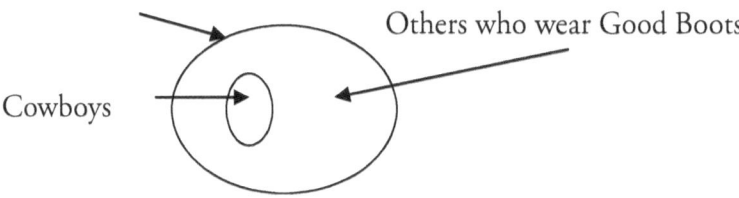

People who wear Good Boots

Others who wear Good Boots

Cowboys

Suggestion: Carefully discuss the missed problems with other students or a Math Teacher. Understanding of the Decision-making problems is very important.

Session 4

The use and misuse of Basic Statistics

Statistical thinking will one day be as necessary for efficient citizenship as the ability to read and write.

H. G. Wells

Rodin's *The Thinker (Golden Gate State, San Francisco, CA,*
(From the Elander File)

A mathematician, like everyone else, lives in the real world. But the objects he works with do not. They live in that other place-the mathematical world.
Something else lives here also. It is called TRUTH.

Jerry P. King
THE ART OF MATHEMATICS

Factors that influence conclusions (valid or invalid) were listed in the Introduction and are repeated below.

Inductive Reasoning
Deductive Reasoning
        Direct and Indirect
What you see or think you do?
What you read and the quality of the material.
What you listen to or don't listen to on the radio, or
        lectures, conversations, or even gossip, and the
        qualifications of the persons providing the information.

31

What you watch on TV and its validity.
Advertisements
Statistics
Polls and their results
> (You need to know how the poll information was collected, when, where, and how it was collected, who was asked, the wording of the questions, plus the number interviewed.)

Statements interpreted as Implications or one of the forms are:
> Converses, Inverses, and Contrapositives

Review 4.1: The Meanings of the Three Averages

The calculating of averages is fairly simple, but interpreting the results accurately can be confusing and is often misleading. You may come across numbers representing averages in the news everyday and as an informed citizen you should be able to understand the correct interpretation of these. You have probably heard the statement: "Figures don't lie, but liars figure." It is important for you to understand the use and misuse of these averages and their implications. This session will help you interpret statements involving statistics and other information that is needed to see the true picture.

The reporting of scores or sets of data to convey the results or the "picture" by using averages is often used. Some examples of these are:

1. The meaning of math scores for a certain grade or class in your school.
2. The average income per family.
3. The average score per game for your school's basketball team.
4. The average stopping distance for a certain make of car.
5. The average pay per employee.
6. The average for your school's ACT or SAT scores.
7. Many economic "pictures" are illustrated by numbers representing averages.

Let's look at the arithmetic average of the following set of scores on an eighth grade math team exam. The team consisted of the five best students.

Scores: 100, 99, 80, 76, 70

The superintendent proudly reported to the newspaper that the average score for the eighth grade 2012 math team is 85. Does this one number give a true picture of the team's scores when no student even scored an 85?

Def: 31: The MEAN or arithmetic average for a set of numbers is the sum of the numbers divided by n, the number of numbers in the set. Formula: Mean = (sum of n scores)/n

Calculate the mean for the above eighth grade math team set of scores. Was the report correct?

The 2011 and 2012 math team scores are listed below. The superintendent reported the teams had the same average in each year.

2012 scores: 100, 99, 80, 76, 70   2011 scores: 100, 95, 91, 80, 59

Calculate the mean score for each year.

In both cases only the average (85) was reported to the parents.
    Was he correct?
Do you think the two teams are the same with regard to their scores?
Which team would you say is better overall? Why?

Should the parents be satisfied with just one number reported for each of the two teams in order to have a ore informative picture? Why?

Now to further confuse the interpretation of data, there are three types of averages. They are the mean, mode and median. Each is used by the business world, in education, the news media, the financial world, and in the medical world. The mean was defined above.

Def: 32: The MODE for a set of data is the most popular or most frequently occurring element or score in the set.

Def: 33: The MEDIAN for a set of data is the middle element or score when the elements or scores are arranged in order from lowest to largest.

What is the median and mode for each of the two math teams listed above? Note: 80 and 91 are the medians, there is no mode.

These three (mean, mode and median) are many times called the average, but the one that is being referred to is not always indicated. Each of these averages will be further clarified and compared in the applications.

Review 4.2: Applications (Some answers are given for your benefit and easy check.)

1. The following is a set of income figures representing the annual payroll in a small business, including the owner's salary.

    $18,000, $18,000, $18,000, $22,000, $25,000, $25,000, $80,000

    Which do you think is the salary of the owner? Why?
    Answer: $80,000

2. Using Definitions 31, 32 and 33 calculate the three averages for the data in 1. Use your calculator.

    The Mean is _____. The Median is _____. The Mode is _____.

    Answers: Mean = $29428.57. Median = $22000. Mode = $18000

3. The mean, mode, and median in #2 were reported as the average!
    a. Which average do you think the owner claimed for the average pay?
        Why?
        Answer: Mean or $29428.57
    b. Which average do you think the employees claimed for the average pay?
        Why?
        Answer: Possibly the mode, $18,000.

4. What is the mode for each of the 2011 and 2012 math team scores?

2012 scores: 100, 99, 80, 76, 70   2011 scores: 100, 95, 91, 80, 59

Answer: There is no mode.

Review 4.3: More Applications with answers

1. In the following calculate and identify the three averages for each of the 5 sets. Use your calculator and/or computer, if needed. First arrange the scores in ascending order, and then calculate the three averages.
   a. 3,5,7   b. 0, 1, 3, 5, 16   c. 0, 1, 1,-2,-3
   d. 2,-3, 6,-7,-1,-4, 7
   e. 71, 65, 98, 57, 64, 69, 87, 88, 79, 48, 77, 80, 50, 75, 85, 90, 99, 30, 94, 96, 80.
   Answers: 1a. Mean = 5, Mode = none, Median = 5
            1b. M = 5, Mode = none, Median = 3
            1c. M-3/5, Mode 1, Median = 1
            1d. M = 0, Mode = none, Median =-1
            1e. M = 1582/21 = 75.3, Mode = 80, Median = 79

2. Many times scores are grouped for easier calculating of the averages and for graphing. In the following case the scores are grouped to intervals of five and assumed the scores are at the center value.

Score	Use	Number of students
96-100	(98)	6 (These 6 scored were actually 96, 97, 98, 99, and 100.)
91-95	(93)	9
86-90	(88)	10
81-85	(83)	13
76-80	(78)	14
71-75	(73)	15
66-70	( )	13 Complete the ( )s
61-65	( )	10
56-60	( )	7
61-65	( )	3

   a. How many students took the test?
   b. What is the sum of all the scores? Assume the scores are at the center of the interval. (6(98) + 9(93) + 10(88) + ... + 3(53)).

   c. What is the mode score?
   d. What is the median score?
   e. What is the mean score?
   f. Construct a graph to show the scores and mark the averages.

Answers: a. 100 b. 7650 c. 73 d. 78 e. 76.5
          f. Graph (The students may bring in a variety of types. One type is started below.)

Intervals	Scores
96-100	******
91-95	*********
86-90	**********
1-85	Complete the graph and indicate the interval
76-80	for the mean, median, and the mode.
71-75	
66-70	
61-65	
56-60	
61-65	***

3. The actual scores which are assigned to each interval in problem 2 are listed below. With a computer or even with some calculators the value for the averages can be easily calculated using the built in programs. There is no need for schools to report only the mean score for a class or school. Actually the whole graph should be reported and the three averages indicated. This is the correct informative way to really compare groups and to detect ways to improve the conclusions.

Student   Scores
   100,98,99,97,97,96
   95,95,94,94,94,93,93,92,91
   86,86,87,87,87,88,88,88,88,89
   85,85,84,84,84,84,84,83,83,83,82,82,81
   80.80.80.79,79,79,78,78,78,77,77,76,76,76
   71,71,71,72,72,72,73,73,73,73,74,74,75,75,75
   66,67,67,67,68,68,68,68,68,68,68,69,69
   61,62,62,62,63,64,64,64,65,65
   56,56,57,57,57,58,60
   51,53,55

a. How many students took the test?
b. What is the sum of all the scores?
c. What is the mode score?
d. What is the median score?
e. What is the mean? (Use your calculator)
    Answers: a. 100 b. 7642 c. Mode 68 d. Median 76
                e. Mean 76.4

4. Montana is a very large state, approximately 700 miles long and
    400 wide, yet the national weather news may assign one number
    to indicate the temperature for the whole state. The change in
    altitude is not even considered. The following numbers indicate
    the variation of August temperatures comparing five of the largest
    cities in Montana.

                    101, 95, 67, 45, 36

a. What is the mean temperature for these readings?
b. What is the median temperature for these readings?
c. What is the mode temperature?
d. Which temperature do you think the National News media
    reported for the State of Montana? Why?
e. What would you report for the state of Montana? Why?
    Answer: There is no one correct figure for the five cities, let
             alone a State temperature.
             a. Mean is 68.8 b. Median is 67. c. There is no mode.
             d and e. Just report the high and low and where these
             are, or for the large cities, or tourist areas.

5. Baseball players keep track of their batting averages. A player has an
    average of 290 after 75 times at bat, what will he have to average in
    the next 40 times at bat to have an average of 300? Be able to justify
    your solution.
    Hint: Does the equation below make any sense to you?
    (Total Hits/Times at bat) = Average
    Answer: 318.75 [75(290) + 40(A)]/115 = 300 Solve or A

6. A baseball player's batting average is 280 in 30 times at bat. How
    many hits will he need in the next 80 times at bat to average of 300
    for the 110 times at bat?
    Estimate the answer before you calculate the correct answer.

Answer: Approximately 30-31 (30.2, but it is impossible to hit a .2) hits.

7. Many colleges and some high schools use A = 4, B = 3, C = 2, D = 1, and F = 0

8. If a student has a 2.0 GPA after 8 courses, then what must the student average in the next 24 courses to have GPA of 3.5. Guess first and then solve using algebra!
   Answer: Average a 4, which means he made all A's.
   Solution hint: {8(2) + 24S}/32 = 3.5

9. Another student at the same school as in #8 has a 2.0 average after 16 courses, then what must she average in the next 16 courses to have a GPA of 3.5. Guess first and then solve for the real answer.
   Answer: It can't be done!! Hint: [16(2) + 16(A)]/32 = 3.5 and solve for A

10. From results of problems 8 and 9, what is your conclusion as to when the high grades should be made?
    Answer: See Summary

11. Calculate the mean weight of the local high school football team's defensive line from last season.

    Challenges:

12. A student drives to the school at 20 mph (rush hour) and drives home at 30 mph. What is the average speed for the round trip?
    Answer: See Summary. This type of question is, almost always, on standardized tests. Hint: Averate rate is Total distance / total time.

13. In the figure below, what is the ratio of the area of the square to the area of the triangle? The side of the square is 6 units. Hints: (Theorems 15 and 16.)

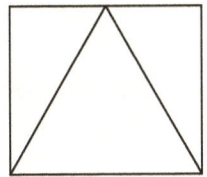

Answer in the Summary

14. Decision making

A school activity advisor said that activity A increased its membership three hundred percent the second year and they had only one member resign. Would you as the superintendent approve of the activity?

a. Yes b. No c. Think about it. d. Ask for more information

Answer in the Summary

15. How (some) people get their information.

The following comparison was recently printed in a newspaper poll. The following was all that was reported as to where each group acquired their news.

	Age groups	
	18-49	50-plus
Newspaper (Daily)	20 %	42%
Online	34%	16%
Television	52%	69 %
Radio	20%	22%
Friends	23%	19%

a. Write a few conclusions.

b. What are some of your questions?

c. What would you add to the Poll to improve the conclusions?

Summary

Def: 31: The MEAN or arithmetic average for a set of numbers is the sum of the numbers divided by n, the number of numbers in the set.
Formula: Mean = (sum of n scores)/n

Def: 32: The MODE for a set of data is the most popular or most frequently occurring element or score in the set.

Def: 33: The MEDIAN for a set of data is the middle element or score when the elements or scores are arranged in order from lowest to largest.

Suggestion: See suggested readings (Index 5) pertaining to statistics.

Answers:

9. Answer: High grades should be earned in the first 2 years!

12. Answer: This question is, almost always, on standardized tests. The answer is 24 mph.
    Hint: Average rate is Distance divided by time or
    Ave Rate = 2d/(d/20 + d/30) = 120/5 = 24.
    Suggest the students discuss this question with their parents and fellow students.

13. Ratio is 2/1 13. d. Ask for more information. (The first year there could have been only one member and the second year 4 members and then one resigned.)

14. a. One conclusion may be that "younger" people use the computer.
    b. One suggestion is to break down each group to smaller groups.
    c. How many were polled, and where, and when, and which stations or TV channels.

Session 5

Applications with Problem Solving

Number rules the Universe
The Pythagoreans

Rodin's *The Thinker (Golden Gate State, San Francisco, CA,*
(From the Elander File)

Rodin's, THE THINKER, Golden Gate State Park, San Francisco,CA

Review 5: Applications with Problem Solving

(We all want to be valid thinkers.)

Do you believe you use mathematics everyday? Well you do, from the time the alarm goes off (numbers) your day is regulated by applications of mathematics. In this session you will discovery a few of these that you may not be familiar with. Organize your solutions and be able to explain them. This is a review and you should work all the exercises even if you think the problem is easy. The answers are in the Summary.

Review 5.1:

1. If you had a rectangular lot 50 ft by 75 ft, how would the area change if you doubled the size to 100 by 150? Write down your

guess as to how much larger, smaller or the same. Than solve for the original area and for the new area. Was your guess correct? (This concept is used in estimating costs by contractors.)
Hint: Index 3: Theorem 31

2. Given a rectangle 3 ft. by 4 ft.
   a. Draw the figure and label it ABCD.
   b. Draw the diagonals and label the point of intersection O.
   c. What is the length of each diagonal?
   d. What is the length of AO and BO?
   Hint: Index 3: Theorem 20, Theorem 12

3. The city owns a park area in the shape of a rectangle, 100 yds by 150 yds.
   Around the inner edge of the park they plan to construct a 4 ft. wide side walking path.
   a. What is the area of the park?
   b. What is the area of the walk in sq. ft?
   c. What is the area of the walk in sq. yds?
   Hint: Index 3: Th. 16

4. A Billboard measures 6 by 10 feet and a larger billboard measures 12 by 20 ft. The employee told the manager the larger one would take 4 qts of paint because the smaller only took 2 qts. The manager disagreed! Who is correct?
   How many qts of paint are needed for the large billboard, based on the area?
   Hint: 3: Th. 31

5. If a circular pizza has a diameter of 9 inches and costs $9.95 and a 12 inch pizza cost $12.95 which is a better buy per square inch? Guess first and then calculate cost per sq. in. of each. (Used in the pizza business)
   Hint: Index 2: Postulate 11

6. Everybody likes pizza but do you ever check the prices for best buy? If a circular pizza has a diameter of 8 inches and costs $6.50, then what should a 12 inch diameter pizza cost, based on area?
   Hint: Calculate the cost per square inch of the small pizza and use

the same cost per square inch to determine the cost of the 12 inch pizza. Can you draw a conclusion from exercises 5 and 6?
Hint: Index 2: Post. 11

7. If you had a triangle shaped garden with a base of 25 ft. and an altitude of 15 ft., how would the area change if:

Altitude

Base

a. You halved the base? Write down your guess and than solve for the original area and the new area. Was your guess correct?
b. If you halved the base and doubled altitude, then what would be the change in the area? Write down your guess and than solve for the original area and the new area. Was your guess correct? (Used in determining the cost of seeds and/or fertilizer.) Can you draw a conclusion? If yes, then write your general conclusion.
Hint: Index 3: Th.15

8. You measured a triangular lot and the sides were 75 ft. by 109 ft. by ** ft.
The **(number) was smeared by rain, but you told the employee the third side is between two measurements. What are the two measurements?
Hint: Index 3: Th.1

9. Draw a 30°-60° right triangle with an altitude from the right angle.

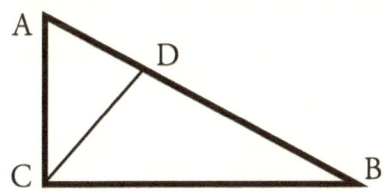

a. Label the largest triangle ABC with C the right angle, and CD the altitude.
b. What is the measure of each of the 6 angles?
c. What is the length of each segment? Given the hypotenuse is 10 units.

Hint: Map the triangles ABC, ACD, BCD
      Index 3: Theorems 20,21,22

10. The ACT test question is:
    a. What is the sum of the interior angles in the following polygon?

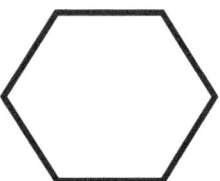

    b. What is the name of this figure? (Used in some college testing programs)

Hint: Index3: Th. 3

11. A student noticed that the number for the circumference of a circle was the same as the number for the area. Can this be? The answer is yes. What is the number?
(Used in college testing) Hint: C = A
Hint: Index 2: Posts. 10, 11

12. This case refers to the skills involved in the games of pool or billiards. The object is to hit the ball so that it will hit side BC and reflects to pocket at A.

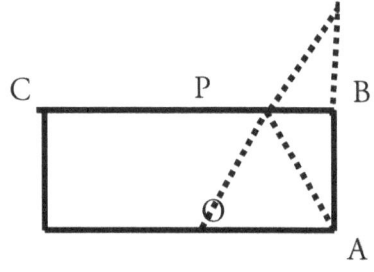

The professional player imagines point T such that AB equals BT and mentally marks point P such that OPT is a straight line. P is the point that he will hit the ball at in order to reflect the ball to A. (Pool tables have marks on the sides to help in determining the point to aim at. Players know their geometry.) Explain why and this method works or better is to try it on a pool table! Hint: Congruent Triangles

13. The park manager wants to build a scenic trapezoid or semicircle background for the stage. This will add to the summer programs both acoustically and artistically. The stage is 45 feet across. The bases for the trapezoid are 45 and 22.5 feet. The altitude for the trapezoid is the same as the radius of the semicircle.

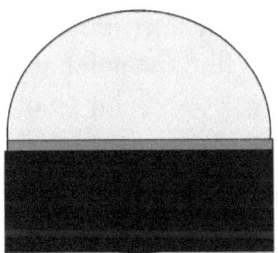 

    a. If the stage is 45 ft. across then what is the area of the semicircle?
    b. What is the area of the Isosceles Trapezoid?
Hint: Index 3: Th. 17

14. Euler's formula for the relationship between the Vertices, Segments, and Regions for polygons is illustrated below.
What is the is the number of Vertices, Segments, and Regions (Inside and outside) in each figure below? Answers are given for the triangle.

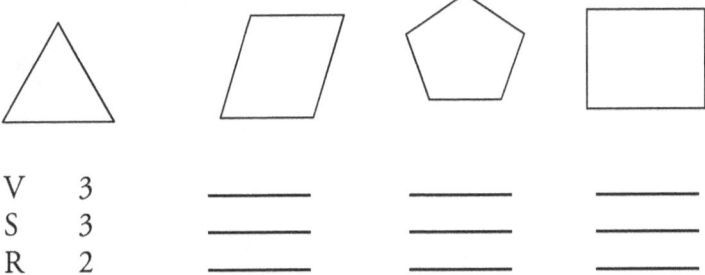

V	3	_____	_____	_____
S	3	_____	_____	_____
R	2	_____	_____	_____

    b. Predict a formula or relationship between R, S, and V.

c. Test your prediction using the figure below. (Inductive reasoning)

15. Draw three different shaped triangles out of stiff uniform cardboard, one acute, one obtuse, and one right. (By stiff is meant non-flexible, and by uniform means same density or thickness throughout.)
    a. With a pencil with a rubber eraser find the point and mark it where the triangles balance on the pencil. (This point is called the Center of Gravity)
    b. Now draw on each triangle, using your ruler, the three medians.
    c. What do you observe as to where the three medians intersect? (Used in engineering) (See Optional Activity following the Summary.)

16. Using the method from number 15, try to locate the Center of Gravity of a non-square quadrilateral cut from the same material used to cut the triangles.
    Test your solution by using the pencil balance test.

17. The Center of Gravity (Also called the sweet spot.) is the point on a baseball bat or a golf club which will result in the longest hits. Ask a coach to explain and demonstrate how to locate the Center of Gravity on a bat or golf club. This also explains why cars tip over when turning at to fast a rate.

18. The following is a sagging gate. How would you recommend it be corrected?

(Used in construction)
Hint: Index 3: Th. 12

19. A carpenter will check to see if a parallelogram is a rectangle by measuring the diagonals. If they are equal, then the parallelogram is a rectangle. Draw a figure and test the validity of the statement. If it is true, try to prove it.
(Used in construction)
Given: Parallelogram ABCD with diagonals AC and BD. AC = BD

Justify: ABCD is a rectangle.

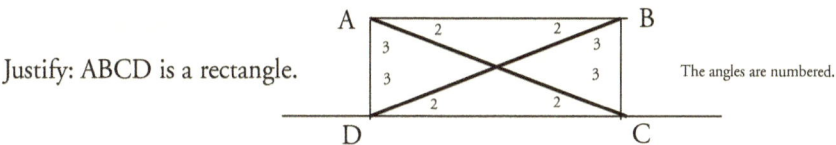

The angles are numbered.

Hints: What do you have to prove for a parallelogram to be a rectangle?
(Prove the angles are right angles or equal to 90 degrees)
Since it is given that ABCD is parallelogram and the diagonals bisect each other, then it must be proved that the figure is a rectangle.
Hint: What makes a parallelogram a rectangle?
Are there any isosceles triangles?
Angles: 2+2+3+3 = 180 degrees. Why?

Hence: 2 + 3 = ? Why?

Therefore . . . .

Write the theorem.

20. What is the sum of the angles in Decagon?

21. Postulate 6 states that three non-collinear points determine a plane. This postulate is used to prove a very practical and very useful theorem.
This Theorem labeled Th. 33 in Index 3. (It could be called the Tripod Theorem or the Telephone Pole Theorem.)

The following figure illustrates the theorem.

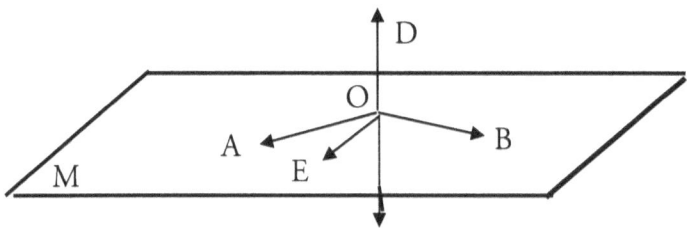

It is given that lines AO and OB are perpendicular to OD. The three points A, O, and B determine the plane M by Post. 2.1. Lines OA and OB are in the plane and perpendicular to line OD. OE is any other line in M that intersects point O and it can be proved that OE is also perpendicular to OD and therefore OD is perpendicular to the plane.

Comment: This useful theorem is used by engineers to erect poles such as flagpoles, utility poles and even the driveway basketball basket poles. All of these should be perpendicular to the ground plane. An example can be seen in your room by looking at the corner where the ceiling intersects two walls.

22. (Indirect reasoning) Four students were called in by the Dean one at a time and asked:
Who wrote the note?

The four students were Geo, Tom, Pete and Joe.
Their statements were:
     Geo: Tom did it.
     Tom: Joe did it.
     Pete: I didn't do it.
     Joe: Tom lied when he said I did it.

The students agreed that only one of the four statements is true.

From their comments, the Dean knew who wrote the note.
Hint: The case below assumes Geo is telling the truth and notice it leads to a contradiction. (Recall that applying indirect

reasoning you assume all possible cases and they all lead to a contradiction but one, then that one is the solution.)

Case 1: If Geo's statement is true, then
    Tom did it.
    Joe did not do it.
    Pete did it.
    Joe did it.

The above does not reveal the person, so in like manner, assume the Tom' statement is true and perhaps that will reveal the guilty person.
Continue in like manner until you know who wrote the note?
Suggestion: Discuss this problem with your parents or friends.

23. Statement: If you a student, then you know your school song.
  Assume the statement is true. Write the following and classify each as valid or not necessarily valid.
  a. Converse
  b. Inverse
  c. Contrapositive
  (The forms of an implication are probably the most misunderstood and misused method than any other decision making by the general public.)

Summary

Write your summary and check your answers.

A very worthwhile Optional Activity:

    Why objects (cars, boats, etc) tip over!
    Use the three different triangles from exercise 15 where the center of gravity has been located and set up the following for a class demonstration to illustrate how the center of gravity point determines also the tipping point.

Method: Take each triangle as illustrated with the equilateral triangle below and tip each (The rotation point is B.), then release and let it fall. When the median (center of gravity, point M) is beyond the vertical line (DB) the triangle will continue to fall so that point A is on the line CB.

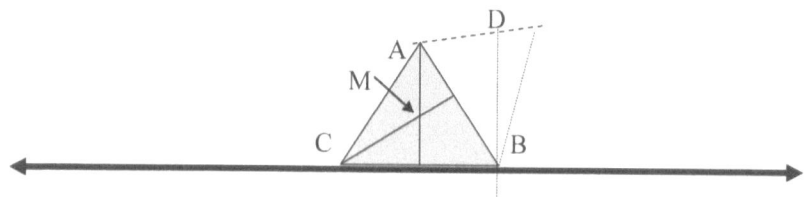

Student can make a large model such that a pin will hold the triangle at point B and this will can easily show the tipping point. (This explains why cars can tip over and boats can roll over.)

Answers:

1. 4 times
2. b. 5 ft.   c. 2.5 ft.
3. a. 15000 sq. yds. b. 5936 sq. ft. c. 660 rounded to the nearest whole sq yd.
   (As a contractor you would have to calculate the number of cubic yards of cement, 4 inches thick, needed for the walk?)
4. The manager since the larger one will take 8 qts. Ratio of areas in 240/60 or 4 times 2 = 8 qts.
5. The 12 inch
6. $13.57 or $13.50 rounded to the half dollar.
7. a. ½ b. Same c. Areas: 796.3 for the semicircle and 795.4 for the trapezoid
8. 184 > ** >34
9. Angles are 30, 60, or 90. Draw the figure on the board.
   Sides are 10, 5, 8.66, 7.5, 4.33, 2.5.
10. a. 720 degrees b. Hexagon
11. R = 2 Hint; C=A or $2\Pi r = \Pi r^2$
12. The shortest distance between two points is a straight line. The angle of reflection equals angle BPT, and PT equals AP.

13. a. Semicircle area is 796.3 sq. ft. b. Triangle is 505.6 sq. ft. Answers may vary due to rounding.

14. a.

V	3	4	5	5
S	3	4	5	8
R	2	2	2	5

   b. V + R = S + 2 Try it for more polygons.
   c. V = 5 S = 11 R = 8

18. Triangle is a rigid figure-insert a diagonal after "squaring" the gate.

19. If the diagonals of a parallelogram are equal, then the parallelogram is a rectangle.

20. 1440 degrees

21. See problem

22. Solution: Pete wrote the note.
       Assume: Joe is telling the truth, then:
           Geo: Tom is Innocent.
           Tom: Joe is innocent.
           Pete: I did it.
           Joe: I am innocent

23. a. If you know the school song, then you are a student. NNV
    b. If you are not a student, then you do not know the school song. NNV
    c. If you do not know the school song, then you are not a student. V

Extra Discovery Activity

Conclusions are made or should be after collecting information from as many creditable sources as one can find. In today's world these sources are TV shows, magazines, books, lectures, radio reports and experts (many times pseudo experts) plus friendly discussions by partially informed and loud people.

The following type of problem requires data organization, observation, plus discovering a conclusion and then testing the discovery. What is the sum of the first 100 even counting numbers? The following is one approach to the solution. (Organization and observation helps to discover the method.)

N	Numbers	sum
1	2	2
2	2+4 =	6
3	2+4+6 =	12
4	2+4+6+8 =	20
5	2+4+6+8+10 =	30
6		

Continue until you see an easy way to arrive at the sum of the first N even numbers. Remember you are looking for an easy method to give the sum of the first 100 even counting numbers.

One answer is $S_e = N(N+1)$. In this case $S_{100e} = 10100$

Session 6

Implications and Polls

I THINK THEREFORE I AM
Rene Descartes

Rodin's *The Thinker (Golden Gate State, San Francisco, CA,*
(From the Elander File)

Review: Logic-Surveys-Conclusions

Pythagoras, the teacher, paid his student three oboli (a coin) for each lesson he attended and noticed that as the weeks passed the boy's initial reluctance to learn was transformed into enthusiasm for knowledge. To test his pupil Pythagoras pretended that he could no longer afford to pay the student and that the lessons would have to stop, at which point the boy offered to pay for his education . . .

Simon Singh
FERMAT'S ENIGMA

This session will illustrate how statements can be better interpreted and will help you make better decisions or conclusions as a result from applying what you have learned in Mathematics.

Sound intriguing?

A branch of mathematics, called LOGIC, and is very useful in decision making. You will always be making decisions in the future, right ones and probably some wrong ones. This Session will just touch on a small bit of this fascinating and very practical topic.

We will begin by investigating statements in a special form.

Review 6.1: IMPLICATIONS

An implication is a statement in the form of If →Then. A few examples from the math world and a few from the everyday world:

> If a triangle has only two equal sides, then it is labeled as Isosceles.
> If a geometric theorem is important, then it should be in the geometry books.
> If the weather forecast is rain, then you should take rain gear to school.
> If students do their homework, then the students will pass the course.
> If you don't abide by the law, then you will be fined.
> If voters are informed on issues, then beneficial laws will be passed.

The base or root word for **implication** is "imply" which provides the meaning for the above examples. Symbolically, this can be written as A → B, where A represents the "If" part and B represents the "then" part.

This is, as you no doubt noticed, a type of short hand or an abbreviated way of writing implications. In other words, A → B will be read "A implies B" or "If A then B." Which translation do you prefer? Most students prefer the second one. This type of sentence is identified as a conditional sentence or statement. You possibly had this in other courses and identified the parts of the A → B as A, the hypothesis, and B as the conclusion. The important implications, proved in geometry are called theorems. In every day experiences they are called conclusions or even laws.

Comment: The United States Declaration of Independence by Jefferson is a great example of A → B. (Have you read it? If not, just read the first 2 pages.)

The question now is: How are implications used and interpreted? For this we will use a diagram, which will show the validity of the implication for the general case. You will have to decide the truth or falsity of the statements. "If A, then B." can be depicted in diagram form as shown below. Remember statements are true or false, but conclusions are valid or invalid.

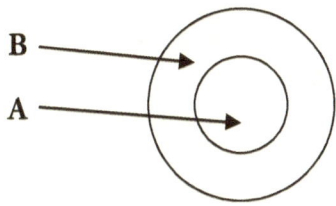

The diagram shows that if you are in circle A, then you are in circle B or A→B. Theorems take this form. Recall from your previous Sessions, the other three forms of an implication theorem are:

Name   Symbolic
Theorem or statement:
                    A → B
        Read: If A, then B.
Converse: B → A
        Read: If B, then A.
Inverse: ~A →~B
        Read: If not A, then not B. The negation symbol, read not, is ~.
Contrapositive: ~B →~A
        Read: If not B, then not A.

Using the diagram below for "If A, then B," classify each form of the implication in the following as valid or may not be valid or maybe.

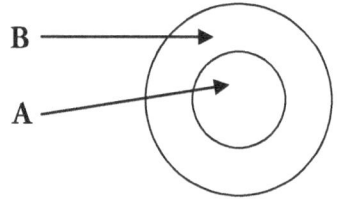

Implication   Name   Your answer
Assume it is valid that **A→B**.   (This is the original statement.)
This means that if in circle A, then in circle B.   Valid

Is it valid that B→A?   (This is the Converse.)   ?
This means that if in circle B, then in circle A.
Is it valid that ~**A** →~**B**?   (This is the Inverse.)   ?
This means that if not in A, then not in B.
Is it valid that ~B→~A?   (This is the Contrapositive.)   ?
This means that if not in B, then not in A.

Do your conclusions agree with the following? Be sure to understand why, if
your answers differ. These are important relationships and are summarized
below! Conclusions are valid or not valid, and statements are true or false.

Theorem   Valid
Converse   Not necessarily valid
Inverse   Not necessarily valid
Contrapositive   valid

Review 6.2: Each statement is written in one of the forms. Write the other
    three forms in the if-then form, and classify each as valid of may be
    valid. (Important, statements are true or false and conclusions are valid
    or not valid)

(Explain your answers as to "T" or "F" or "could be" and validity for the
    following.)

1. Statement: If a person is a teacher, then that person enjoys students.
    (Classified as True and valid)
        The first three forms are answered for you.
    Write the Converse.

Answer: If a person enjoys students, then the person is a teacher.

(Could be true or false since non-teachers may also enjoy students. Conclusion not always valid)

Write the Inverse.

Answer: If a person is a not a teacher, then the person doesn't enjoy students.

(Could be true or false -why? Conclusion not always valid)

Write the Contrapositive.

Answer: If a person does not enjoy students, then that person is not a teacher. True and conclusion is valid.

2. Defintion: If the figure is a triangle, then it is a polygon. (True and valid)

    Write the Converse.

    Answer: If the figure is a polygon, then it is a triangle.
    Could be F and invalid

    Write the Inverse.

    Answer: If the figure is not a triangle, then it is not a polygon.
    Could be F and invalid.

    Write the Contrapositive.

    Answer: If the figure is not a polygon, then it is not a triangle.
    True and valid

3. Statement: If you own property, then you pay taxes.
    (True and valid)
    (Explain your answers.)

    Write the Converse.

    Answer: If you pay taxes, then you own property.
    Could be F and not always valid.

    Write the Inverse.

    Answer: If you do not own property, then you do not pay taxes.
    Could be F and invalid.

    Write the Contrapositive.

    Answer: If you do not pay taxes, then you do not own property.
    True and valid

4. Statement: If you live in the Montana, then you live in the United States of America. (True)
    (Explain your answers.)

Write the Converse.

> Answer: If you live in the U. S. of A, then you live in Montana. Not necessary valid or true.

Write the Inverse.

> Answer: If you don't live in the MT, then you don't live in the U.S. of A. Not necessary valid or true.

Write the Contrapositive.

> Answer: If you don't live in the U. S. of A, then you don't live in Montana. Valid and T

5. Statement: If you display the flag on the 4th of July, then you are a patriot.

> (Assume True and valid)
>
> (Explain your answers.)

Write the Converse.

> Answer: If you are a patriot, then you display the flag on the 4th of July. Not necessary true and invalid.

Write the Inverse.

> Answer: If you don't display the flag on the 4th of July, you are not a patriot. Not necessary true and invalid.

Write the Contrapositive.

> Answer: If you are not a patriot, then you don't display the flag on the 4th of July. True and valid

6. Statement: If you wear the Flag Pin, then you are patriotic.

> (Assume valid and true.) It will help you to understand this situation by using the diagram below.

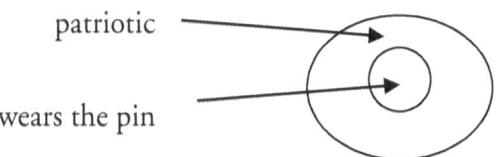

(This forms of this statement, by some groups in the 2004 election actually interpreted as true and valid for all the forms and persons were falsely accused.)

Explain your answers. This is valuable for a group discussion.

Write the Converse.
    Answer:
Write the Inverse.
    Answer:
Write the Contrapositive.
    Answer

Keep in mind that statements are true or false, and conclusions are valid or invalid. This important relationship will become more meaningful as you have more experience.

7. Given: Law: If you are a high school graduate, then you are Educated, according to some States.
      Note: This also may bring up the role of definition.
    (You classify as True or False and Explain your answers.)
  Write the Converse.
      Answer: If you are educated, according to your State, then you are a high school graduate.
  Write the Inverse.
      Answer: If you are not a high school graduate, then you are not educated according to your State.
  Write the Contrapositive.
      Answer: If you are not educated according to your State, then you are not a high school graduate.

8. Given: Ad: If you use brand X, then you will be popular.
      (False, but ads would like you to believe their statements.)
      (Explain your answers.)
  Write the Converse.
      Answer: If you are popular, then you use brand X.
  Write the Inverse.
      Answer: If you don't use brand x, then you will not be popular.
  Write the Contrapositive.
      Answer: If you are not popular, then you don't use brand X.

9. An Interpretation of an Ad:
      A professional golfer uses ball W, then you should use it also for a better game. (The ad wants you to think this is true.)
      (Classify each as to T or F or could be and as to validity.)

Write the Converse.

Answer: For a better game, I should use ball W.

Write the Inverse.

Answer: If I don't use ball W, I won't have a better game.

Write the Contrapositive.

Answer: If I don't want a better game, then don't use ball W.

10. Statement: If wages are increased, then there will be inflation.

(Do you think the statement is true or false? Explain your answers, which will depend on how you answered the question.)

Write the Converse.

Answer: If there is inflation, then wages were increased.

Write the Inverse.

Answer: If wages are not increased, then there will not be inflation.

Write the Contrapositive.

Answer: If there is no inflation, then wages are not increased.

Recall in one of the previous Sessions the word validity was used as well as True or False. Conclusions are valid or invalid and statements are true or false. A conclusion may be valid, but the statement can be false.

Example: A State reported that there are very few accidents involving cars traveling at speeds over 100 MPH, therefore one could conclude that limits should permit drivers to drive at these high rates. The conclusion is valid, but false for several reasons. List some.

Review 6.3: S is the given statement and C is the conclusion. Classify the conclusions as valid or invalid and in some cases also as true or false. Answers are highlighted. It will help you to arrive at the correct answer, if you draw the circle diagrams for the problem.

1. S1: If you're a teacher, then you enjoy students.

S2: My uncle is a teacher.

C: My uncle enjoys students.

Explanation: The conclusion is valid since from S1 the circle of teachers is within the circle of persons who enjoy students and since the uncle is a teacher, then he is in the circle of persons who enjoy students. See diagram below.

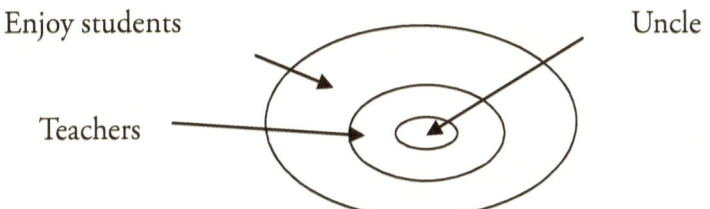

Enjoy students

Uncle

Teachers

2. S1: If you are a teacher, then you enjoy students.
    S2: My uncle enjoys students.
    C: My uncle is a teacher.
        Converse form Invalid? Why?

    Note: Answers are in blue.

3. S1: If the figure is a triangle, then it has three sides.
    S2: All three sided figures are polygons.
    Conclusion: All triangles are polygons.
        Valid or invalid? C is true but S2 is false
        This a three sided figure, but not a triangle.

    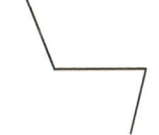

4. Given: Let S be "Seth is smart."
            Let: C be "Chelsea is cute."
            Let: E be "Eric is clever."
    Translate the following:
        a. S implies C. b. E →C. c. ~E →C. d. ~S →~C

5. Given: Let S1 be "Judges are intelligent."
            Let S2 be "Ben is a judge."
            Conclusion: "Ben is intelligent."
                Valid or invalid? Why?

6. Given: Let S1 be "Judges are intelligent."
            Let S2 be "Ben is intelligent."
            Conclusion: "Ben is judge."
                Valid or invalid? Why?

7. Given: Let S1 be "Judges are intelligent."
            Let S2 be "Ben is not a judge."

Conclusion: "Ben is not intelligence."
Valid or invalid? Why?

8. Given: a. If you are a high school graduate, then you are educated according to your State.
   b. Pete is a high school drop out.
   Conclusion: Pete is not educated.
   Valid or invalid? Why? Inverse
   But may be True by the State's Def.

9. Given: 1. If you are a high school graduate, then you are educated according to your State.
   2. Pete is not educated.
   Conclusion: Pete is not a high school graduate.
   Valid or invalid? Why? Contrapositive
   True by State's Def.

10. Given: a. If you don't wear the Flag Pin, then you are not patriotic.
    b. John did not wear a Flag Pin.
    Conclusion: John is not patriotic.
    Depends on the T or F of "a", probably false.
    (T or F in ?)

11. Given: a. If you wear the Flag Pin, then you are patriotic. (Assume true)
    b. John did wear a Flag Pin.
    Conclusion: John is patriotic.
    Invalid or valid. Why? The assumption

12. Given: a. If you don't wear the Flag Pin, then you are not patriotic. (not true)
    b. John did wear a Flag Pin.
    Your conclusion? John is patriotic.
    Which form of the statement a is b? (Inverse, converse, or contrapositive)
    True

Note: Check a few ads from magazines and write each ad as an implication and the four forms.

Review 6.4: Surveys and a "touch" of Statistics

This section will provide an insight into common procedures that many times lead to questionable conclusions. This section investigates information data collected from test scores or samples of data, like your GPA (grade point average). Once the data is collected, a few numbers are usually calculated like mean, mode, median, range, to provide a description of the data. (These terms and how to use them may have been partially explained in others courses.) Naturally, the most accurate description would be to use the entire set of data to tell the story, but if the data is very large, like the total population of a city or State, then usually only a sample is selected. In this section, we will discuss some pros and cons of collecting the data. Statisticians use the phrase "random selection" when referring to how the data is collected. Randomness means the selection is haphazard, indiscriminate, accidental, uncontrolled, or hopefully simply non-biased.

An example of biased sampling occurred in the 1936 presidential election between Franklin D. Roosevelt and Al Landon. The Literary Digest sampled a set of voters via telephone listings and automobile registrations. They interpreted the data that resulted in a prediction for a Republican landslide, and Al Landon's election. The Republicans discovered after the count was in that many who voted did not have telephones or own cars, but knew the way to the voting booth, resulting in FDR's election. The statisticians had not polled a random sample!

Surveys are done for socio-political reasons to provide part of the "picture" attempting to give the impression that the prediction or outcome is based on a scientific method of inquiry using mathematics. Listed below are the common "weaknesses" associated with polling? Polls are use to predict an outcome by using some of the weaknesses in the collection process.

1. Arriving at conclusions based on:
   a. Too small a sample for the conclusion

Example: A school publishes the SAT scores for its seniors as an indicator of quality education when only a small percentage of the students may take the test.

b. A random sample that is not random with reference to whom and where x is.

Example: Nielson ratings for the popularity of TV shows are based on a sample of approximately 1200 families out of over 100 million viewers. Recently a survey polled 808 voters at random and then predicted which presidential candidate was favored to win.

c. Not providing sufficient information to correctly arrive at conclusions.

Example: One number, such as the mean, does not give a sufficient picture. Two very different sets can have the same mean.

d. Questions the pollsters ask may be worded to encourage a biased answer, usually a yes or no response.

Example: Are you for a pollution free environment?
(Who wouldn't be!)

e. Advertisements will contain undefined terms, or associations

Example: A car manufacturer may claim a certain car is rated BEST in its class without stating who rated it or what other cars are in the same class, if any.

f. Polls do not indicate what the selected people know about the subject.

Example: Are you in favor of good television? (Do "you" know what is good T.V.? Do they define "good TV?")

g. Quotes or recommendations made by experts or pseudo experts to sway the public opinion and the poll results.

Example: An actor in a sitcom who plays the part of a doctor and then he or she recommends a product in an ad.

2. How can survey reports be improved or more meaningful?
   a. State the size of the sample.
   b. Clarify how, when, and where the data was collected.
      (Telephone, state the time of day for calls, mail, computer, interview, and where.)
   c. State the questions exactly as they were worded, with key words defined?

    d. If the term average is used, clarify which average is used (mean, mode, or median).

    e. Show a summary of the complete set of data.

Review 6.5: Advertising and Surveys (Listen to the students' answers and comments. Require them to defend their answers.)

1. Professional basketball player J uses brand S shoe as shown in a TV ad. (Weakness 1.e.)
   a. What is the objective of the ad?
   b. What is the implied translation?
   c. Is the translation in b valid?

2. If you were to poll first the student body and then the faculty, predict and explain your answers or results to the following statements.
   a. Students and faculty should have an hour lunch period.
   b. Students should be dismissed early to attend all athletic events.
   c. Students should be in classes for a full 180 days per year.
   d. All final grades in all classes shall fit the normal curve. (What is a Normal Curve? (Role of definition))
   e. Do you think the responses would be considered bias with regard to various groups?
      (Be able to justify your answers.)

3. The following statement appeared as a local newspaper headline. "The minimum wage increased last year and also smoking among teenagers."
   a. What is one possible false conclusion?
   b. What possibly is the implied remedy for teenage smoking?

4. The news media reports that most auto accidents occur within five miles from home.
   Write a possible conclusion?

5. Reports: How national surveys collect data.
   a. The Nielsen ratings, Gallup surveys, and over 50 other professional organizations.
   b. Data collected by E-mail, telephones, mailing lists, personal questions.

6. Collect a variety of graphs from newspapers and/or magazines and discuss their meanings or obvious conclusions.

7. On a certain bill in Congress the Republicans said the people were against it and the Democrats said the people were for it. They both quoted poll results. How could this be? What questions would you ask or the media should have asked?

8. Write in if→then form what you feel the following ad is trying to communicate?

    An actor on TV played the role of a Doctor and gave a very persuasive presentation for the following. (He also knew the importance of what is called perfect numbers.)

One 12 oz. bottle per day plus 6 minute of fast walking is the formula for a NEW LIFE.

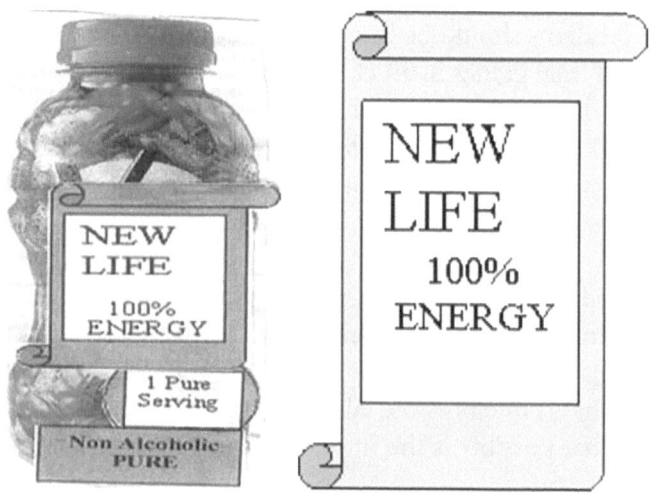

Give NEW LIFE a 28 day trial test for a new life!
You won't be sorry!

a. What are the perfect numbers?
b. Which words need defining? (like PURE)
c. What are some questions you would like answered?

9. Write a summary for Session 6.

Suggested reading: This book by Huff is very interesting.
   HOW TO LIE WITH STATISTICS by Darrell Huff

Summary

Terms and forms of an implication
Name   Symbolic form

Theorem or statement:

	A $\rightarrow$ B	Read: If A, then B.
Converse:	B $\rightarrow$ A	Read: If B, then A.
Inverse:	~A $\rightarrow$ ~B	Read: If not A, then not B.

The negation symbol, read not, is ~.

| Contrapositive: | ~B $\rightarrow$ ~A | Read: If not B, then not A. |

The issue of Validity and Truth for conclusions. Statement and Contrapositive are always valid, then Converse and Inverse are not valid.

The weaknesses of polls, ads, and data collection.

Session 7

3-D Applications

Students of mathematics . . . the first time something new is studied seem hopelessly confused . . . Then, upon returning (to the concept) after a rest, . . . everything has fallen into place.

E. T. Bell
MEN OF MATHEMATICS

Rodin's *The Thinker (Golden Gate State, San Francisco, CA,*
(From the Elander File)

We live in a three-dimensional world, so to solve 3-D problems involving mathematics is a logical activity. You will encounter these types of problems sooner or later. In other words, these applications are practical. Geometry Essentials is used as the basis for the review for College Entrance Exams for the simple reason that Geometry and Critical Thinking has been proven to be the applicants' weakest areas from the results of past entrance tests.

To solve the problems use your notes, session Summaries, calculator, and your ingenuity. If you have a question, then ask for help. Who do you ask? Anybody who you think can help you. A person can learn a lot by asking questions. It will help you to understand the problem if you sketch a figure that you think represents the problem. Organize your work so others can follow your method!

Review 7.1: 3-D Applications
        Hints: Theorems 38, 39, 40.

1. Sphere A has a radius of 2 ft. and sphere B has a radius of 6 ft.
    a. What is the area of sphere A?
    b. What is the area of sphere B?
    c. What is the ratio of the areas of sphere B to sphere A?
    d. What is the ratio of radii of sphere B to sphere A?
    e. Write a general conclusion pertaining to the ratio of the areas
        for two spheres with regard to the radii.
    f. If the ratio of the radii is x/y, then the ratio of their surface areas
        is ?/?.

2. Sphere A has a radius of 2 units and sphere B has a radius of 8
    units.
    a. What is the volume of sphere A?
    b. What is the volume of sphere B?
    c. What is the ratio of the volumes, B to A?
    d. What is the ratio of the two radii of spheres B to A?
    e. Write a general conclusion pertaining to the ratio of the volumes
        for two spheres related to the ratio of their radii.
    f. If the ratio of the radii is a/b, then the ratio of their volumes is
        ?/?.

3. Two engineers each designed a container that holds 8 cubic feet.
    One container was a sphere and the other a cube. Which container
    has the least amount of surface area? Guess first and then calculate
    the answers. (This type of packaging problem is used in retail
    sales.)
    Hint: Solve for the measure of the side of the cube and also the
        radius of the sphere, and then calculate the surface area of each.
        Theorem 39

4. A tank truck consists of a cylinder and two hemispheres, one
    hemisphere at each end of the cylinder for aerodynamic reasons.
    The diameter is 8 feet and the total length of the tank is 45 feet.
    Use pi as 3.14.
    Hint: Theorems 34, 38

    a. Draw the figure that represents the tank.
    b. What is radius of the hemisphere?
    c. What is the length of the cylinder?
    d. What is the volume of the cylinder?
    e. What is the volume of one of the hemispheres?
    f. What is the volume of the tank?
    g. If a cubic foot holds 7.54 gallons, then how many gallons are in the tank?
    h. Assume the tank is hauling gas and the wholesale price is $3.05 per gal, then what is the value of the gas in the tank? Suggestion: Use the current price of gas.

Answers: Review 7.1 (Answers may vary due to the value used for pi.)

1. a. $Aa = 4(\pi)4 = 16\pi = 50.24$ sq. units
    b. $Ab = 4(\pi)36 = 144\pi = 447.84$ sq. units
    c. $Ab/Aa = 144/16 = 9/1$
    d. $Rb/Ra = 3/1$
    e. and f.: The ratio of the areas of two spheres is the square of the ratio of their radii. $a/A = (r/R)^2$ This is Th. 40.

2. a. $Va = (4/3) \pi 8 = 32\pi/3 = 33.51$ cu. units
    b. $Vb = (4/3) \pi 512 = 2048\pi = 2144.77$ cu. units
    c. $Vb/Va = 2044.77/(33.51) = 64/1$
    d. $Rb/Ra = 64/1$
    e. and f.:The ratio of the volumes of two spheres is the cube of the ratio of their radii. $v/V = (r/R)^3$

3. Area of the sphere is 19.33 sq. units.
    Area of the cube is 24sq. units. The sphere has less surface area.

4. a. Drawing b. 4 ft c. 37 d. 1859.82 cu. ft. e. 134.04 cu. ft.
    f. 2127.90 cu. ft. g. 16044.37 gal. h. $48935.33

Review 7.2: More applications (Answers depend on the value used for pi.)

1. What is the measure of the radius of a ball, if the measure of the surface area is the same as the measure of the volume?

2. If a double-dip ice cream cone consists of two hemispheres and the diameter of each is 2 inches, then:
   a. What is the volume of the ice cream?
   b. How many double dip cones can be made (in theory) from a half gallon container of ice cream that measures 7x5x4 inches?
   c. If a half gallon of ice cream costs $4.25 and a double dip cone costs 2.50, then what is the profit from the cones from the half gallon of ice cream?

3. Which is a better buy, if both paper cups of soft drink are the same price? A cylinder, with a radius of one inch and is 4 inches high, or one that has a 1.5 inch radius and is 2 inches high?
   Hint: Draw the two glasses to scale and guess your answer before doing the calculations.

4. A yard is 50 by 70 feet. If the rainfall was measured at 1 inch, then how many cubic feet of water fell on the yard?

5. Convert your answer to number 4 to the nearest whole number of gallons. (One cubic foot contains 7.54 gallons.)

Answers: Review 7.2 (Answers may vary depending on value used for pi.)

1. R = 3 units
2. a. 4.18 cu. in. b. 33.49 cones c. 2.5(33)-4.25 = $78.25
3. The cup with the 1.5 radius is the better buy.
   One inch cup contains 12.56 cu. in. and the 1.5 in. cup contains 14.13 cu. inches.
4. 291.67 cu. ft. 5. 2199 gals

Activity 7.3: More applications

1. Archimedes favorite problem
   a. Draw a cylinder with radius R and a height of 2R.
   b. Draw a sphere with radius R.
   c. Draw a cone with radius R and height of 2R.
   d. Draw the figure showing the cone and sphere inside the cylinder.
   e. Write the formula for the volume of each.

$$V_{sph} = \underline{\hspace{1cm}} \quad V_{cone} = \underline{\hspace{1cm}} \quad V_{cyl} = \underline{\hspace{1cm}}$$

f. What are the ratios?

$$V_{cyl}/V_{cone} = \underline{}?/? \quad V_{cyl}/V_{sph} = ?/? \quad V_{sph}/V_{cone} = ?/?$$

Answers: 3/1 3/2 2/1

g. The volume of the cylinder is ____ times the volume of the cone.

h. The volume of the sphere is ____times the volume of the sphere.

Note: Notice that the Cylinder is 3 times the cone, and the sphere is 2 times the cone. This was Archimedes' favorite theorem and the drawing of the cylinder with the inscribed sphere and cone was on his tomb.

2. It is reported that Lewis and Clark were very impressed with the Bull boats the Mandan Native Americans used for transportation on the Missouri River. These boats resembled wooden framed hemispheres covered with buffalo skins. Assume one is about 6 feet in diameter and a cubic foot of water weigh approximately 60 pounds. Archimedes discovered the buoyancy principle: That an object in water is supported by an amount equal to the weight of the water the object displaces. Could a Bull boat carry 1500 pounds?

Note: Stephenie Ambrose Tubbs: **The Lewis and Clark Companion.** Henry Holt and Company, *page* 31

The way to get rich in 27 days.

3. How to become a millionaire in 27 days! (In a 31 day month the banks may be closed 4 days.)

a. Either use a calendar or make a list similar to the following.

The first day you put a penny in the bank. The second day you put in 2 cents. The third day you put in 4 cents and the 4th day 8 cents, etc. A is the amount deposited and S is the sum of the deposits. D is the day.

Complete table for the 27 days. Look for a PATTERN then make predictions for the amounts on the 15th and 27th days.

Day
1. A = 1 S= 1
2. A = 2 S= 3
3. A = 4 S= 7
4. A =___ S=___

b. What is the problem with this method of becoming a millionaire?

c. Check your answer for the total by the following formula. This is one formula for the answer. Your pattern method may reveal an easier one!
The math Formula is: $S = (-1 + r^n)$ where S is the Sum and r is 2, and n is 15 or 27.

4. If a rectangle has sides measuring 12 and 5, then what is the measure of the diagonals?

5. If a circle has diameter of 6, then what is the area of the circle?

6. If a right triangle has sides of 3, 4, 5 and the midpoints of the sides are connected, then what is the area of the internal triangle formed by the three new line segments?

7. If the sides of a triangle are 5, 15 and x, then what is the value of x if the square root of x is an integer?

8. If two diameters of a circle (radius is 6) are perpendicular and then the endpoints are connected, what is the area of the square that is formed?

9. Many times people will take a short cut across a corner of a lot or lawn. In the picture below, what is the distance saved by taking the short cut?
    Given: AB and AC are each 25 feet.

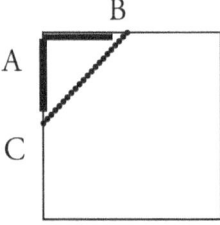

a. What is the length of CB
b. What is the length of AB + AC?
c. How much shorter is BC than AB + AC?

d. Should the owner be concerned?

10. Recall that conclusions are valid or invalid, and statements are true or false.
Consider the following ad stated in If→then form.

If it is a Q-Computer, then it is quality-built.

a. Which of the following are **valid** conclusions and why?
1. If the computer is quality-built, then it is a Q-Computer
2. If the computer is not quality-built, then it is not a Q-Computer.
3. If the computer is not a Q-Computer, then it is not quality-built
b. What does the ad want you to believe is true?

Answers: Review 7.3

1. a-d. Drawings
e. Vsp = $4\pi r^3/3$ Vco = $2\pi r^3/3$ Vcyl = $2\pi r^3$
f. Vcyl/Vco = 3/1 Vcyl/ Vsp = 3/2 Vsp/Vco = 2/1
g. 3 h. 2
2. Vol. is 37.58 cu. ft. and this supports more than 1500 lbs
3. c. You have to be rich to make the last few payments.
The math formula is: S = $(-1 + r^n)$ = When n is 15, S is 327.67, and when n is 27, then S is $1,342,177.27. Some students may recognize that the S is 2^D-1.
4. 13
5. 28.6
6. 1.5
7. 16

8. 72 sq. units
9. a. 35-36 ft. b. 50 ft. c. 14-15 ft. but the lawn is ruined.
10. a.   Statement 2 is valid, but probably not true.
    b.   "If you want quality, then buy a Q-Computer." or "If you don't
         buy a Q-Computer, then you won't get quality."

Summary

Th.40: The ratio of the areas of two spheres is the square of the ratio of
their radii. $a/A = (r/R)^2$

Th. 41: The ratio of the volumes of two spheres is the cube of the ratio
of their radii. $v/V = (r/R)^3$

Archimedes' favorite theorem is the ratio of the volumes of the cylinder
to the sphere to the cone is 3 to 2 to 1 when inscribed.

Session 8

That they (all citizens) might excel in public discussions on philosophic or scientific questions, they must be educated(rhetoric, philosophy, mathematics, and astronomy).
The Athenian Sophist School Curriculum(480 B.C.E.)
F. Cajorie

Rodin's *The Thinker (Golden Gate State, San Francisco, CA,*
(From the Elander File)

Review 8.1: Applications

Note: The answers are provided for your benefit. Be able to justify your answers. Remember the "proof of the pudding is in the eating!" Many of these questions involve the content related to the questions on the math portion of the college entrance exams, plus Decision Making Skills. Use your notes and calculator, when needed! Make a note of the questions you miss and the related theorems.

There will be no hints for the problems in this review, but use the indices if needed. It is better to take your time and get the problem correct than to "race" through them and have the wrong answers. Which do you think your future employer prefers?

1. What is the sum of the angles of a plane triangle?

2. If a theorem has been proved, then is the converse always valid?

3. If the sides of a triangle are 24, 16 and x, then what does this information tell you about the length of side x?

4. a. Do any three points determine a triangle?
   b. How many points are needed to determine a geometric plane?

5. If a square has a side of 10 ft then:
   a. What is the perimeter?
   b. What is the area?
   c. What is the measure of the diagonal? (Give answer and to the nearest inch)

6. If the vertex angle of an isosceles triangle is 72 degrees, then what is the measure of the other angles?

7. What is the method used for arriving at a conclusion by indirect proof?

8. If the figure is a triangle, then it is a polygon? Is the converse true?

9. Can a median of a triangle ever be outside the triangle? Explain.

10. Can an altitude of a triangle ever be outside the triangle? Explain.

11. If a ray divides an angle into two equal angles, then what is the name of the ray?

12. What is a postulate?

13. What is the test for a valid definition?

14. If two lines intersect, then the ___ angles are equal.

15. In the following figure, what is the measure of angle ACD?

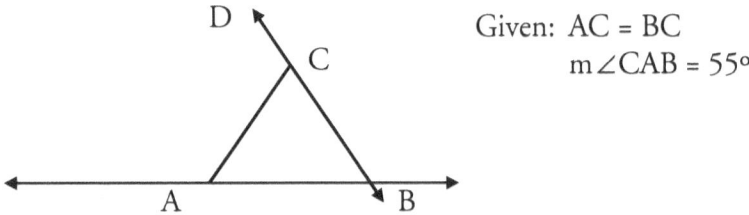

Given: AC = BC
m∠CAB = 55°

16. All logical and illogical conclusions are derived from, ___, ___, ___ and ___.

17. Similar figures in geometry have two properties, ? and ?.

18. In the following figure:

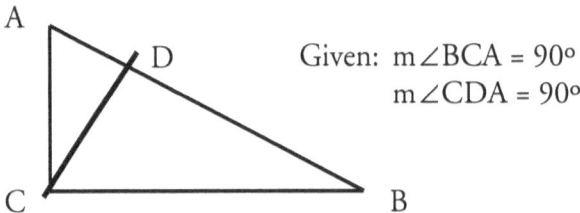

Given: m∠BCA = 90°
m∠CDA = 90°

   a. Map the three similar triangles.
   b. If angle B is 30 degrees and AB is 10, then calculate the measures of the other five line segments to the nearest tenth. (AC, CD, CB, AD, DB.)
   c. What is the ratio of the sides of the large triangle to the sides of the smallest triangle?

19. A round pizza has a diameter of 14 inches.
   a. What is its area to the nearest sq. in.?
   b. If it is put into a square box, 1.5 inches high for delivery, then what is the volume of the box? Assume the pizza just fits in the box.
   c. Draw the layout picture of the box, and calculate the area of the pizza box.

20. What is the volume of a cone and pyramid each with an altitude of 8 inches and the circumference of the base of the cone is 31.42 inches.

The pyramid has a square base with a perimeter of 36 inches? (Answers to the nearest two decimal places)

21. A large circular clock reads 4 o'clock, what is the measure of smaller angle formed by the hands? Answer in degrees.

22. The clock in number 21 is in a 10-inch square wooden frame.

   a. What is the measure of the radius?
   b. The vertex of the angle formed by the hands is at the __ of the __.
   c. The sides of the square are ___ to the circle.
   d. What is the area of the circle? (Nearest tenth)
   e. What is the area of the square?
   f. What is the area of the square outside the circle? (Nearest tenth)
   g. What is the degree measure of smaller angle when the hands read 8 o'clock?

23. What is the **Total Exterior** area to the nearest sq. in. of a pyramid with a square base (perimeter of 60 inches) and a height of 8 in?

24. If a sphere (globe) has a diameter of 20 inches, then: (Use pi equal to 3.14)
   a. Draw the 3-D view of the globe showing the equator and diameter AOB, where O is the center. Draw a radius perpendicular to AB, label it ON.
   b. What is the volume of the sphere?
   c. What is the area of the sphere?
   d. What is the circumference of the circle of latitude on the globe with center 6 inches from O on ON?

25. A 40-inch line segment (AT) is tangent to a circle with a 30 inch radius at point T. The center of the circle is labeled O. What is the length of the segment AO?

26. In a ghost town in Montana your Dad found a map under a rock near the Clark Fork River. The map read: The gold is located at the center of the circle determined by:
    a. A pine tree with "T" carved on its trunk.
    b. A large rock that points North.
    c. The entrance to the cave.

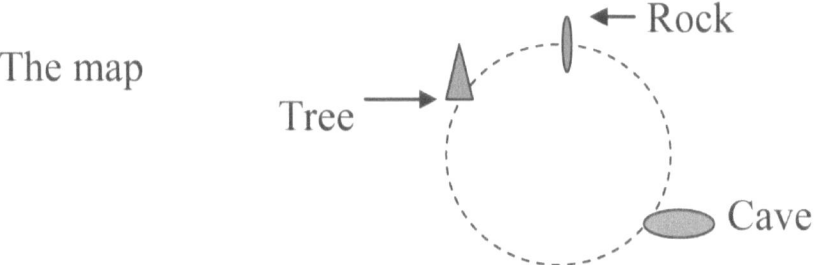

The map

Explain how you, mathematically, would find the center and the gold.

27. Draw the three-dimensional object given the following views.

28. Is statement "b" valid given statement "a" is True ? Why?
    a. If Nick does not follow directions, then he will be fired.
    b. If Nick follows directions, then he will not be fired.

29. Complete: Conclusions are ___ or not ___, but statements are ___ or ___.

30. Complete: All conclusions are based on ___, ___, ___, and ___.

31. United States has a Republic form of government where the people elect the officials to run the government. In order to have an efficient government the following are needed (among others):
   a. An educated well informed voter.
   b. Well informed elected representatives.
   c. News media that covers all sides of the issues.
The news media has a huge responsibility to keep the citizens informed on all sides of the issues. Informed citizens depend on that responsibility.
Statement: If you are an informed citizen, then the news media covers all sides of the issues.
   Assume the above statement to be valid.
Which of the following are valid?
   a. If you have a news media that covers all sides of the issues, then you will have informed citizens.
   b. If the citizens are not informed, then the news media does not cover all sides of the issues.
   c. If the news media does not cover all sides of the issues, then the citizens will not be informed.
32. What is the sum of the interior angles in the following figure?

Answers

1. 180 degrees 2. No
3. 8 < x < 40 4. a. No b. Three non-collinear
5. a. P = 40 ft. b. A = 100 sq. ft. 2 in. c. D =10√2 or 14 ft.
6. 54 degrees
7. List all the possibilities and assume all but one. If all but one of the assumed possibilities prove false or lead to a contradiction, then the one possibility left is the correct one.
8. Yes, No, 9. No

10. Yes 11. Angle bisector
12. A statement assumed to be true is an assumption.
13. Is the definition true when reversed?
14. Opposite or vertical angles are equal. 15. 110 degrees
16. Undefined, defined, assumptions or postulates, and proven theorems or laws that follow from the first three.
17. Corresponding angles are equal in measure and corresponding sides are in equal ratios.
18. a

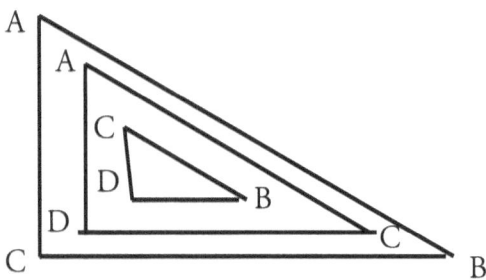

   b.   AB = 10 CB =8.6 or 5√3 AC = 5 AD = 2.5
        CD = 5√3/2 or 4.3
        DB = 7.5
   c.   Ratio is 2/1
19. a. 153.9 + sq. in. b. 294 + cu. in. c. 476 sq. in.

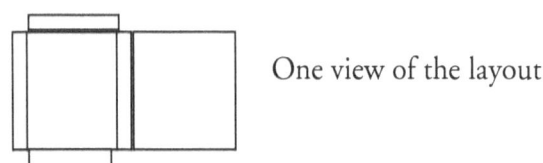

One view of the layout.

20. Vp = 216 cu. in. Vc = 209 cu. in.
21. 120 degrees
22. a. 5 inches b. Center of circle
   c.   Tangent d. 78.54 sq. inches
   e.   100 sq. inches f. 21.5 sq. inches g. 120 degrees
23. A_T = 553.8 sq, in.

24. a. Drawing

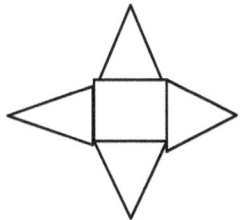

    b. V = 41886.7 cu. in. c. A = 1256 sq. in.

    d. C = 50.27in. Answers may vary depending on the value of pi
        that is used.

25. 50 inches

26. Draw the chords and construct the perpendicular bisectors. The
intersection of the bisectors is the center.

27. Drawing.

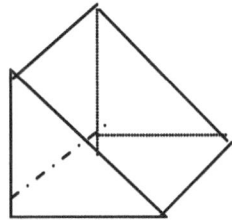

28. No, since "b" is the inverse of "a."

30. Undefined terms, defined terms, assumptions and Theorems or
Laws that follow from the first three. This is a repeat of number
16, but very important!

31. a. If you have a news media that covers all sides of the issues, then
you will have informed citizens. Invalid, since you may not read or
listen to the media. (Converse case)

    b. If the citizens are not informed, then the news media does not
        cover all sides of the issues. Invalid (The inverse case)

    c. (answer) If the news media does not cover all sides of the issues,
        then the citizens will not be informed. Valid (Contrapositive
        case)

Note: The following diagram may help to understand the conclusion
for a, b, and c.

NM 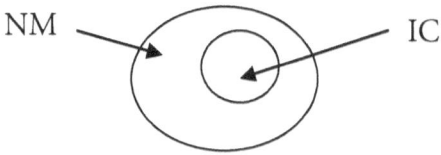 IC

32. 720 degrees

Session 9

Algebra and Number Activities
(Some different types of thinking!)

Mathematics uncovers the "mystery" behind a problem.

Rodin's *The Thinker (Golden Gate State, San Francisco, CA,*
(From the Elander File)

RODIN'S, THE THINKER, Golden State Park, San Francisco, CA

NUMBER THEORY—ALGEBRA—
REASONING ACTIVITIES POTPOURRI

A set of activities to provide review for solving different types of unique problems. You may even wish to try some of these on your friends or parents. This of course means you will have to really understand the solutions! These types of activities were acquired by reading interesting books (see Bibliography), attending professional meetings and trying these types in the classroom. The results impressed the author and he recognized their potential value, hence the desire to share them with others. Many former students responded positively.

Some answers and hints to the Activities are given on the last page.

Activity 1:　A problem with a predetermined answer!
　　　　　　Which areas of study are important for the future?
　　　　　　You know that students are always asking, Why they need to
　　　　　　know a subject or what will the benefits be?"
　　　　　　This Activity will help a student determine which subject will
　　　　　　help the most in preparation for college or vocational school.
　　　　　　(If it were only this easy.)

　　　　　　You should select the most beneficial subject in your opinion
　　　　　　and use the corresponding number to your guess.

1. English
2. World History
3. Biology
4. Physics
5. U.S. History
6. Chemistry
7. Athletics
8. Activities
9. Mathematics

　　　　　　After your selection as to your opinion, then perform the
　　　　　　following steps that will reveal the subject that provides the
　　　　　　best preparation.

1. Select your guess as to the subject area and multiply the corresponding counting number by 3.
2. Add 3 to your answer in step 1.
3. Multiply the number from step 2 by 9.
4. Add the digits in the answer to step 3 until you have a one-digit number.
5. Match your answer to step 4 with the corresponding subject or program. This is the subject area most valuable for you.

Comment:　The answer will always come out 9, indicating Math is the
　　　　　　most valuable! This answer is prearranged based on what is
　　　　　　called **Casting out Nines**. Which was taught at one time

in the grade schools before calculators and should now be used (the authors opinion) in checking solutions to addition, subtraction, multiplication, or division problems.

Example: 7539 + 8642 = 16181 Now add the digits in the three numbers resulting in 24, 20, and 17. Now add the 20 and 24, which is 44. Now add the two digits of 44 to give the single digit 8. Do the same with 17. (The sum of the digits in the answer.) and the sum is 8. Since the two numbers (7539 and 8642) which were added give 8 and the original answer (16181) also gives 8 you know the answer 16181 is correct. (This method of checking works for the four basic operations.) Check to see if this problem has the correct answer

Is the sum correct? Use the casting out nines method to check?
        147 + 258 + 963 = 1368

Activity 2.   This intriguing problem creates the mystery as to how a problem can tell the right or wrong solution.

1. Write the first 9 counting numbers and circle the one you write the poorest.
2. Multiply the circled digit by 9.
3. Multiply the answer in step 3 by the following, 12345679. (no calculator) You should recognize the answer and why it is the right answer!
4. The correct answer will contain only the circled digit. If you don't get the Poorest Digit for the answer, then do the multiplication again.

Activity 3:   How to become a millionaire in 30 days. Sound interesting! (This problem can create the how curiosity in your friends, which you can explain.)
        Method: Use a 30 day month calendar and mark in the first day square to indicate one penny, Also in square 1 put a 1 to indicate the total sum. The square for day 2 should contain a 2 (for 2 cents) and a 3 for the total sum. Now complete the next few days with the amounts until you see a pattern and can answer the amounts for the 30th day.

(Some students prefer the following table to help them see the pattern.)

Day	Deposit	Total Amount
1	1	1
2	2	3
3	4	7
4	8	15
5	16	31
?	?	?

Comment: Each deposit is a power of 2
          Each Total Amount is a power of ? and ? (Your answer)
          How is the answer related to the Day?
          Then arrive at the Total Amount for the 30days, hence a millionaire.
          (The 30 day Total Amount is $10,737,418.23. Use your calculator)

Activity 4. How large is the college class?
             Many times in college the lecture class size is larger than in high school.
             Professor IQ reported 1/5 of his students earned A's, ¼ earned B's, ½ earned C's, and 3 students failed. How large was the lecture class?

         Answer: 60 students

Activity 5. How many of each are purchased?
             A teacher purchased the following for use in her classes. Colored magic markers at $5 each, pens at $4 each, and pencils at 10 for a dollar. The total number of items was 100 and oddly the total cost was $100. How many of each are purchased?

         Answer: 12 markers, 8 pens, and 80 pencils

Comment: If you try to solve this using algebra, it turns out to be a Diaphontine equation, which you may think is impossible.

These equations have more variables than equations, but the other condition or restriction is the requirement of counting numbers for answers.

Can you think of a few other conditions, like the number of markers is less than what number? This will narrow down the possibilities.

These conditions are: Can you list a few other few known conditions?

Counting numbers only

There are less than 100 pens.

Activity 6. What number?

Henry challenged the class with the following problem. "I am thinking of a counting number less than 100 and if you divide it by 3, or 4, or 2 the remainder is 1, but if you divide by 5 the remainder is 0. What are the numbers?" One student said the number ends in a ? or ?

a. What are numbers the ?s marks?

b. What is the number?

Activity 7. Dad, please send money!

A college student, as you would expect, often emails home asking for more money. This student knew his parents enjoyed little tricks or puzzles, so he emailed the following puzzle request. Using only the ten digits representing letters he sent the following message. (Only one digit for a letter, example if **E** is 5 then no other letter can be 5, but if there are two **E's** then both would be 5.) The message was:

$$\begin{array}{r} \text{SEND} \\ + \text{MORE} \\ \hline \text{MONEY} \end{array}$$

The father placed the decimal point in the answer between the N and E. Why?

How much money will the father send?

a. What digit is the M? Why?

b. What can D or E not be?

c. How much is the student requesting?

Activity 8.   Clock Arithmetic
There are two cases to be considered, the 12-hour clock or the 24-hour clock.
Case 1. The 12-hour clocks which requires a.m. or p.m. to tell the time of day.

1. 2 + 4 =?   2. 3+ 6 =?   3. 4+ 6 = ?   4. 6+ 8 = ?
Answers: 6   9   10   2
5. 8 - 9 = ?   6. 10 + 11 = ?7. 3 - 8 = ?   8. 3 x 8 = ?
Answers: 11   9   7   3

Case 2. The 24-hour clock (Military time)

1. 2 + 4 = ?   2. 3+16 = ?   3. 14+16 = ?   4. 16+ 23 = ?
Answers: 6   19   6   39 = 15

5. 8 - 9 = ?   6. 10 + 11 = ?7. 3 - 8 = ?   8. 3 x 9 = ?
Answers: 23   21   19   3
Do you understand the answers?

Suggestion: Draw a round clock face for each time type to help solve the problems and demonstrate the answers.

Activity 9.   1=0?? How can this be?
The symbol (∞) in the following represents a very large number.

Case 1. 1/∞ = 0 or 1 = 0 (∞) or 0, which now reads 1 = 0.
How can this be?
Case 2. ∞/1 = 0 or ∞ = 1(0) or 0, which now reads a very large number equals 0.
How can this be?

Activity 10. Average speed (SAT or ACT question)
A person average 20 mph driving to work in the rush hour. Leaves early and averages 30 mph diving home. What is the average MPH for the round trip?
Answer: a. 25 mph, the average of 20 and 30.
b.   23.6 mph since it took longer at 20 mph

      c. 24.6 mph since it took less than 25.

      d. 26.4 mph since the home trip is faster.

      e. 24mph

Comment: Ask your friends and parents for the their answers and then explain that the correct answer is 24 mph.

Many will select the wrong answer (25) and their curiosity as to why will be surprising.

Activity 11. Summer pay

Pete and John both had summer jobs for June, July, and August that paid the same amount the first month. But at the end of June, Pete was given a 10% raise and John had to take a decrease of 10%. Ironically, on August 1st Pete was given a 10% cut and John was given a 10% increase.

Questions
1. What did each student make during the summer, if the June pay was $100??
2. What did each student make in the month of August?
3. What did each student make during the summer, if the pay was $P?

Activity 12. Diagonal of a 4th-dimensional cube from a mathematical viewpoint.
a. Draw a square of side s and calculate the length of the diagonal.

    Answer:
b. Draw a cube of side of s and calculate the length of the cube's diagonal.

    Answer:
c. By inductive reasoning the diagonal of a 4th dimensional cube is ?

    Asnwer:
d. Predict the length of the diagonal of a 5th dimensional cube.

    Answer:

Comment: Check a dictionary for the word Tesseract. (You may have to go to the Oxford Dictionary.

Activity 13. How many lanes is the coach planning for?

The college track coach suggested to the budget committee that the college build a mile circular track and add 200feet to the circumference for the width of the track. The committee doubted the 200 feet would be enough.

What do you think? In the drawing below the track is the gray area.

If a lane is about 3 feet wide, then how many lanes is the coach planning for?

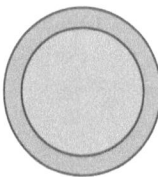

The answer is one of the following, 2, 4, 6, 8, 10, 12. (Justify your answer.)

Activity 14. Saturday's pay

Three high school freshmen sold ice cream bars at Saturdays Game and decided to share the tips based on how many bars each sold. The pay per game was $15 plus tips. The total tips came to $14.95. If Joe sold 2/3 as many as Art and Fred sold ¾ as many as Art, then how much did each earn in tips?

Activity 15. If you take medicine C, then you will be cured of decease D. (Valid)
   a. If you are cured of D, then you took Med C.
   b. If you aren't cured of D, then you didn't take Med C.
   c. If you didn't take med C, then you won't be cured of D.
   d. Classify each of the above as valid or invalid

Activity 16. A mental induction problem.
   a. What do you estimate that $(1 + 1/n)$ equals when n is a very large integer?

b. What do you estimate that $(1 + 1/n)^n$ equals when n is a very large integer?

c. Now evaluate b using your calculator as n is 1000, 5000, 10000?

Activity 17. The Bankers' Rule of Seventy-two should be understood by all. It is one of the most important, miss understood, and harmful financial factors in our society. You will need a scientific calculator for this activity. The conclusion is based on Inductive Reasoning.

The Bankers" Rule of 72 is: Divide 72 by the rate of interest and the answer is the number of years it takes for the money to approximately double.

Case 1.   If the rate of interest is 9%, then investment will double in 8 years.

Case 2.   You invest $1000 at the rate of 8%, then in 9 years the amount is approximately $2000.

In the above 2 examples the rate of interest is per year and the time period is per year. The case where this is really harmful is used by credit card companies. Where the rate and time periods is in months,

Case 3.   Credit card company "Serve you" charges 18% interest per month on the amount of money that you don't payoff each month. If you failed to pay the total amount owed, how many months would it take for the owed amount to double? Answer 72/18 = 4 months

Case 4.   Some credit card companies charge 30% rate (or higher) per month, which means when a person who could not payoff the debt on time would double the debt in ____ months

$A = P(1 + r)^y$ is the formula. The formula reads as A (the amount of money the investment or debt has grown to) is equal to the investment (or debt) times $(1 + r\%)^t$, where r is the rate of interest per t (t and r are in the same time periods). (What number is % equal?)

Example: In case 3 above the r and t are in months, therefore;

A = 1(1 + .24)^4 = (use your calculator) 2.4 or approximately 2 rounded off.

Case 5. What rate does your credit card (or your parent's) company charge? What is the doubling time period?

18. A map of a triangular lot has the measurements 75 by 125 by X. The X measurement has been burred but you know it is greater than ? and less than?

19. (Pattern Type) Complete the next row of numbers in the following:

$$
\begin{array}{ccccccc}
& & & 1 & & & \\
& & 1 & & 1 & & \\
& & 1 & 2 & 1 & & \\
& 1 & 3 & 3 & 1 & & \\
1 & & 4 & 6 & 4 & & 1 \\
\end{array}
$$

— — — — — —

20. What is the sum of the interior angles in this figure?

21. Why does a 4 legged chair sometimes wobble?

22. If 2 points will determine one line segment, then how many segments could be drawn using the following 4 points? (5 points?), (6 points?), (n points?)

Create a table and look for a pattern.

Points	Segments
2	1

23. How many words in the following statement are not definable?
Statement: Now is the time for all good men to come to the aid of their country.

24. Is this a valid definition?
A dog is a 4 legged animal.

25. In right triangle ABC, what is the measure of AB if
AC = 4 and CB = 3?

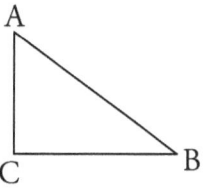

26. All congruent triangles are similar, but all similar triangles are not congruent. True or false?

27. Are these two triangle similar?
Triangle A has sides of 4, 7and 5.
Triangle B has side of 10, 8, and 14.

28. The gate in the following figure is sagging. What would you add to fix it?

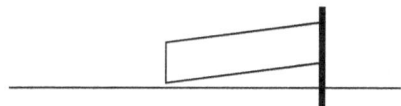

29. If each letter represents a unique digit, then what digit is B equal to in the following addition problem? What digit can T not be?

$$\begin{array}{r} HIT \\ +\ HIT \\ \hline BALL \end{array}$$

30. In the following figure, how would you tell or know if the flagpole is perpendicular to the plane (ground).

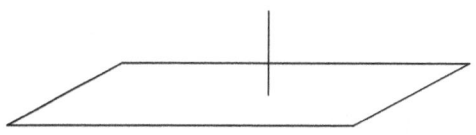

31. What is a theorem?

32. What is a postulate?

33. What is the volume of each of the following figures?
 a. Box L = 10 W = 4 and H = 6

 b. Cylinder: R = 5
      H = 10

 c. A baseball with a 4 inch diameter.

34. If we assume the following statement is valid, then which of the following (a, b, or c) do you think is valid?
 Statement: If students do their homework, then they will pass.
 a. If students pass, then they did their homework.
 b. If students do not do their homework, then they will not pass.
 c. If students do not pass, then they did not do their homework

35. Grades from two small classes in the same subject and teacher.

Class	am	pm
A's	5	3
B's	4	6
C's	1	2
D's	1	1

 a. Draw a graph, from the data, for each class.
 b. Answer the following(c, d, and e) from the graphs.
 c. Calculate the mean, mode, median, and GPA for each class.
 d. The superintendent said the two were equal since they have the same average. Was he right about the averages (mean)?
 e. Which do you think is the better class?

Answers to activites with some explanations and hints.

1. and 2 see activity
3. The pattern is Total Amount = 2^D-1 and then convert to dollars.
4. See activity
5. Diaphontine equation problem

Conditions: Markers + 4P(pens) + p(pencils) = 100 or p= 100-M-P

5M + 4P + p/10 = \$100 or p= 1000-50M-40P

M < 20, Pens < 100and pencils <100

Algebra: 5M + 4P + p/10 =100 (Money equation)

4M + 4P + 4p = 400 (Number of items)

Subtracting: M +0P-39p/10 =-300 or M =-300 +39p/10

You know that 39p/10 > 300 in order for M to be positive and p < then 100 and is a multiple of 10. Hence only 90 or 80 equals p and M< 20. Subbing the two possible p values results in M=51 or 12 and 51 I to large. This also means P is 8. Answer: M = 12, Pens = 8 and pencils = 80. Q.E.D.

6. Answers: a. 5 0r 0 b. 25 or 85 (Method ?)

7. a. 1 b. 0 c. 9567 + 1085 = 10652 or \$106.52

8. Answer is given-see activity.

9. ∞ is not a number.

10. Solution: Average speed is D/T or and the distance is D (one way), then the average is Total distance divided by total time or D/20 +D/30.

Ave. mph = 2D/(D/20 +D/30) or 24mph

11. Answers:
   1. Pete: 100 + 110 + 99 = \$309 John: 100 + 90 + 99 = \$289
   2. Pete made \$99 and John made \$99.
   3. Pete made \$3.09P and John made \$2.89P.

12. a. D of a square is D = s$\sqrt{2}$. b. D for a cube is s$\sqrt{3}$ c. s$\sqrt{4}$ or 2s
   By inductive reasoning an n-dimensional cube would have a diagonal equal to s$\sqrt{n}$.

13. Justification: $2\pi RL - 2\pi RT = 200$

                    $RL - RT = 200\ 2\varpi$

                    Track width is 31.8 feet or 10 full lanes.

14. Answers. Art $6.19, Joe $4.13, Fred $4.65

15. d. Statement a is invalid.
     Statement b is valid
     Statement c is invalid.

16. $(1 + 1/n)^n$ approaches the math. constant e (2.718145927 . . . .).

17. see activity

18. $50 < x < 200$ 19. 1 5 10 10 5 1

20. 360 degrees 21. 4 planes

22. 6 segments

Points	Segments
2	1
3	3
4	6

23. Undefined is between 6 and 9. The two that probably need defining are "good" and "aid."

24. No 25. 5 units 26. True 27. Yes

28. Lift the gate to form a rectangle and then add a diagonal.

29. One (1)

30. Check for perpendicularity from two directions. See theorem index.

31. A proven important mathematical statement.

32. An Assumption

33. P is 40 units Area (sq) is 100 sq. units
Circumference (circle) 10(3.14) = 31.4 units
Area (circle) is 25 (3.14) = 78.5 sq. units

33. a. Vol. of 240 cu. units
    b. Vol. of cylinder is 785 cu. units.
    c. Vol. of ball is 33.49 cu. units

34. "c" is valid.

35. a.

Grade	Am	Pm
A	*****	***
B	****	******
C	*	**
D	*	*

    b.

Mean	A	B
Mode	B	B
Median	B	B

    c.

Calculator	Am	Pm
Mean	3.5	3.5
Mode	A	B
Median	B	B
GPA	3.2	2.9

    d. Basically, YES.
    e. Explain your answer!

Now wait one week and then work Session 10.

Session 10

Epilogue or The Final Word

The important thing is to not stop questioning...
Albert Einstein

Rodin's *The Thinker (Golden Gate State, San Francisco, CA,*
(From the Elander File)

Epilogue and Post-Test

Congratulations, you have completed a review that will improve your test scores and Decision Making skills for Basic High School Math skills and a few Critical Thinking or Decision Making topics. (Basic Geometry and basic logic has been reported to be the poorest results on college entrance exams, hence basic Decision Making Skills plus some Algebra was incorporated into this Review program. These sessions helped you understand the key theorems in 2-D and 3-D Geometry, plus to provide some worthwhile experiences in thinking, not only in the areas of applied mathematics, but in the world of Everyday Decision Making Skills. All thinkers should understand that conclusions are based on undefined terms, defined terms, assumptions, and theorems or previously arrived at decisions such as laws. You should now understand why Plato had posted at his school's entrance:

LET NO MAN
IGNORANT OF GEOMETRY ENTER HERE.

Note: Plato, as you now understand, did not mean just the study of geometric points, lines and planes, plus geometric figures, but he meant the training in logical thinking obtained from the study of a logically developed geometry. Also his school or academy was for male adults, who were going to be the leaders in their society

Permit me to close with the statement, which should be the objective of every high school.

LET NO STUDENT
IGNORANT OF GEOMETRY EXIT HERE.

(This result depends on how the course is taught.)

Keep in mind that you have only been exposed to the "tip" of the mathematical iceberg. Like the visible iceberg is only a small portion of it, there is a vast amount of math still uncovered. Perhaps you will take another course in the future. A Democracy requires a decision making populace and as some person once said: Freedom requires a lot of responsibility. I sincerely hope you can say that this review was worthwhile, even enjoyable at times, and helped you become a better thinker. See the Bibliography for some very interesting books, which hopefully, you will read a few.

Respectfully,
Jim Elander (Author)
Retired in beautiful Montana
Missoula, MT
Email: *jelander@aol.com* or elanderje@gmail.com

Now take this post-exam and calculate your percentage correct score. What would you expect your percentage to be? _____%

Post-Exam

Use you notes and calculator. No time limit! There is no guesswork in Math, you either know the problem or you don't. Your employer will not pay for guesswork.

1. A map of a triangular lot has the measurements 75 by 125 by X. The X measurement has been burred but you know it is greater than ? and less than ?.

2. (Pattern Type) Complete the next row of numbers in the following:

$$
\begin{array}{ccccc}
 & & 1 & & \\
 & & 1\ 1 & & \\
 & & 2\ 2\ 1 & & \\
 & 2\ 3\ 3\ 1 & & \\
1\ & 4\ & 6\ & 4\ & 1 \\
\end{array}
$$

__ __ __ __ __ __

3. What is the sum of the angles in this figure?

4. Why does a 4-legged chair sometimes wobble?

5. If 2 points will determine one line segment, then how many segments could be drawn using the following 6 points?

6. How many words in the following statement are not definable? Statement: Now is the time for all good men to come to the aid of their country.

7. Is this a valid definition?
A dog is a 4-legged animal.

8. In right triangle ABC, what is the measure of AB if
AC = 4 and CB = 3?

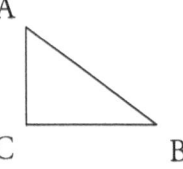

9. All congruent triangles are similar, but all similar triangles are not congruent. True or false?

10. Are these two triangles similar?
Triangle A has sides of 4, 7and 5.
Triangle B has side of 10, 8, and 14.

11. The gate in the following figure is sagging. What would you add to fix it?

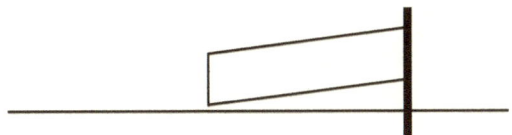

12. If each letter represents a unique digit, then what digit is B equal to in the following addition problem? What digit can T not be? (Can you solve the addition problem?)

$$\begin{array}{r} H\,I\,T \\ +\underline{\;H\,I\,T} \\ B\,A\,L\,L \end{array}$$

13. In the following figure, how would you tell or know if the flagpole is perpendicular to the plane (ground).

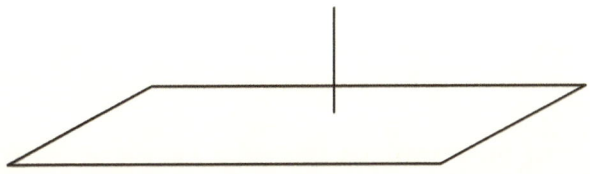

14. What is a theorem?

15. What is a postulate?

16. In the following figure:  A circle is inside a square.
The side of the square is 10
   a. What is the perimeter of the square?
   b. What is the area of the square?
   c. What is the circumference of the circle?
   d. What is the area of the circle?

17. What is the volume of each of the following figures?
   a. Box L = 10 W = 4 and H = 6

   b. Cylinder: R = 5
              H = 10

   c. A baseball with a 4 inch diameter.

18. If we assume the following statement is valid, then which of the following (a, b, or c) do you think must be valid?

Statement: If students do their homework, then they will pass.
   a. If students pass, then they did their homework.
   b. If students do not do their homework, then they will not pass.
   c. If students do not pass, then they did not do their homework

19. What is the median, the mode, and the mean for the following set of scores?
$$(5, 8, 12, 9, 12, 9)$$

20. Which of these two classes would you say is better from these scores?

	Class AM:	Class PM:
A's	5	3
B's	4	6
C's	1	2
D's	1	1

   a. Draw a graph, from the data, for each class.

Answer the following (b,c, and e) from the graphs or mathematics.

b. Calculate the mean, mode, median, and which average is the GPA. class.

c. The superintendent said the two were equal since they have the same average. Was he right about the averages (mean)?

d. Which do you think is the better class?

Answers to Post Exam: Number of answers is 33. Compare your score with the test score on the Entrance Exam. You should be satisfied with the improvement!

1. $50 < x < 200$ 2. 1 5 10 10 5 1
3. 360 degrees 4. 4 planes
5. 15 segments
6. Undefined is between 6 and 9. The two that probably need defining are "good" and "aid."
7. No 8. 5 units 9. True 10. Yes
11. Lift the gate to form a rectangle and then add a diagonal.
12. One (1) One solution is $672 + 672 = 1344$.)
13. Check for perpendicularity from two directions
14. A proven important mathematical statement.
15. An Assumption
16. P is 40 units Area (sq) is 100 sq. units
   Circumference (circle) $10(3.14) = 31.4$ units
   Area (circle) is $25 (3.14) = 78.5$ sq. units
17. a. Vol. of 240 cu. units
   b. Vol. of cylinder is 785 cu. units.
   c. Vol. of ball is 33.49 cu. units
18. "c" is valid. (Contrapositive, a. and b maybe valid.)
19. Median is 9. Mode is 9. Mean is 8.6.
20. a. graphs (This is one type of graph.)

AM	PM
A *****	A ***
B ****	B ******
C *	C **
D *	C *

b. ? (your answer as to why)   c.   AM   PM   d. yes
                                  Mean: 3.5   3.5
                                  Mode   A   B
                                  Mediam B   B
                                  GPA   3.2   3.2

Your score is _____ right out of 33. What is your percent correct score?
Compare it with what you expected. Are you satisfied?

Session 11

Additional Selected Activities
for Math Review

(Many of these activities are related to types of problems on entrance exams.)

Suggestion: Try to relate these activities to practical applications common to your students' back ground or the community to make the activities more relevant.

Comment: A teacher should not tell their students what to think, but teach them HOW to think.

1. **A proof for (-1)(-1) = +1:** (Many students are just told to memorize that the product of 2 negatives is positive, but really need to have a justification.)
   Proof:
   (-1)(-1) equals (?) The real question!
   (-1)(1) =-1 1 is the identity element or number.
   (-1) (0) = 0 Property of zero.
   (-1) [1 + (-1)] = 0 why?
   (-1)[(1+(-!)] or (-1)(1) +(-1)(-1) = 0 Distributive property.
           but-1 + (-1)(-1) = 0 therefore (-1)(-1) = +1 QED.

Comment: % means 1/100, this evolved into 0/0. (You can see the two zeros and the fraction bar.)

2. **Voting interpretations:** If 75% of the eligible voters actually voted, then 1 out of how many did not vote?
   Solution: 75% voted or 75/100 so 25% or 25/100 did not vote, which is reduced to 1/4 or 1 out of 4.

3. **What age is represented by this numeral?** The wise guy in the class said today is his √32√8 birthday. How old is he?
   Answer: √32 is 4√2 and √8 is 2√2 or the age is 16.

4. **Indirect reasoning solves the problem.**
   Some money was missing from the clubs treasury. Only 4 students had access to the funds. Only one of the 4 is telling the truth. From the following which one is the truth?

Student	Comment
A.	I didn't steal the MONEY.
B.	I was lying.
C.	B is lying.
D.	B stole the MONEY

   Answer: B told the truth.

Comment: (You may have to ask for help on Number 5.)

5. **Related to lotteries: Students need lots of practice with this type of activity!**
   What is the probability of tossing a 3 with one toss of a die?
   a. Ask the class to guess first and explain their guess? Answer: 1/6
   b. Now ask their guess to toss two threes with a pair of dice? Ask for an explanation. Answer: 1/36
   c. What is the probability of tossing a 1 or (not both) a 3 with a pair of dice?
      Let them explain their answers. Answer: $2\{(5/6)\ (1/6) + (5/6)\ (1/6)\} = 20/36$ or 5/9
   d. What is the probability of a tossing a 3 and a 1 with a pair of dice?
      Answer: $(1/6)(1/6) = 1/36$ (Could be a different interpretation.)

Comment: Be sure to explain c and d and the meaning of OR and AND! There are two kinds of probability, Mathematical or Empirical. Use a pair of dice to illustrate the empirical with various cases.

6. **Understanding braking!** If a person is traveling 70 MPH in a new car and it takes 1/4 th of a second to apply the brakes, then:

a. How far does the car travel in 1 second? Answer: 103ft nearest whole number.

b. Now it should be easy to determine the distance in ¼ second. Answer: 26ft

c. How far does a car traveling at 60 mph travel in 1 minute? Answer: 1 mile

7. **ACT Test questions**:

a. What is the complement of an angle whose supplement is 120 degrees?
Answer: 10 degrees

b. The midpoint of a 6 inch chord in a circle is 4 inches from the center of the circle. What is the circumference of the circle in terms of pi?
Answer: 10π

c. Two cones have the same circular base area, but the altitude of one is 8 inches and the altitude of the smaller cone is 4 inches. If the small cone filled with ice tea cost 20 cents, then what should the large cone cost?
Answer: Not less than 40 cents

d. Two circular cones, the large cone has a base with a radius of 2 inches and an altitude of 4 inches. The smaller cone has a base radius of 1 inch and an altitude of 2 inches. What is the radio of their volumes?
Answer: VL/VS = 8/1

e. What values for X will make this equation valid.
$|3x+15| = 45$
Answer: +10 or -20

f. What value can x not be in the following expression?
(50X-33)/ (2.5X - 5)
X ≠ 2 (can't divide by 0)

Integer Problems

8. **a.** What integer meets the following three conditions?
1. The number is less than 50.
2. The number is a multiple of 5.
3. The number is a perfect square.
Answer: 25

b. What are the missing numbers in the following sequence?
3,9,7,13,11,?,15,?
Answer: 17, 21

9. Which is larger, $(1/2)^2$ or 2^{-2}? Answer: They are equal.

10. What is the 2 digit prime factor of 52? Answer: 13

11. 120 is the sum of the first x counting numbers. What integer is X?
Method of attack for the solution is the key. 1+ 2 =?, 1+2+3 =? Look for the pattern! Answer: 15

12. a. What is the number for the value of a googol?
b. What is the largest power of 10 on your calculator? Answer: (10^{99})

Problems related to Geometry

13. What is the area of this isosceles trapezoid?

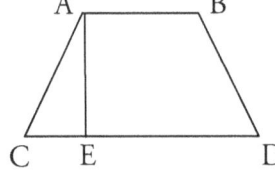

Given: AC is 5, AB is 7
Angle CEA is 90 degrees and AE is 4.

Answer: 40sq. units

14. If a regular polygon has exterior angles of 18 degrees, then how many sides does it have? Answer: 20

15. What is the ratio of the volume of cone in figure 1 to the sum of the volumes in Figure 2? First guess then solve for the ratio. The two cylinders are congruent.

Figure 1 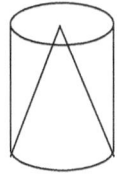   Figure 2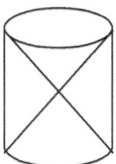

Answer: The ratio is 1/1

16. a. What is the slope of any line parallel to this line, 3x-2y=15?
       Answer: 3/2
    b. What is the x-intercept for the given line? (5,0)
    c. What is the y-intercept for the given line? (0,7.5)

Business

17. A dealer said any high school student could by brand X shoes for this price?
    The $50 pair of shoes were priced at (6-8÷2)4x3-1) for a student. Was it a good buy and what is the price? Answer: yes, $22

18. **Display problem:** What is the height of this figure, if each circle has a diameter of 10? Answer: 27units

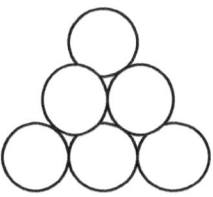

19. Jim, Bud and Greg started a summer business, where Jim put in $500, Bud $1500, and Greg put in as much as Jim and Bud together. If the business had a profit of $2400. How much should each get if the profit is shared according to the investments amounts? Answer: $300, $900, $1200

19. What is the measure of the shortest side in the following figure?

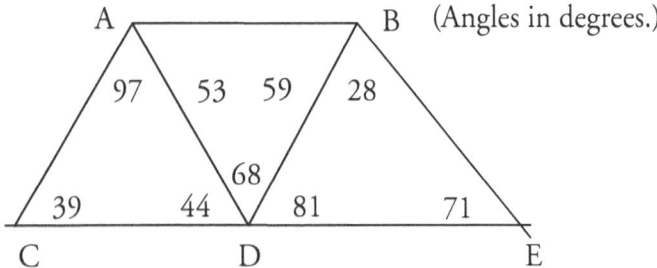

(Angles in degrees.)

Hint: The figure is not drawn to scale! What theorem is this conclusion based on.? Answer: DE

A few general problems arriving at conclusions

20. The school was mailing out notices for future events and asked 32 students to come in on Saturday to prepare the notices. They asked how long the chore would take? The teacher said it was known that 4 students can do 400 notices in 1 hour. The teacher said less than 3 hours. It actually took only 2 hours.
    How many notices did they prepare? Answer: 6400.

21. In 2009 the local post office handled 5,000 letters per day on the average for the first 10 days. The rest of the month the P. O. handled 200,000 letters.
    What did the P.O. handle on the average per day for the rest of the month of April. Answer: 10,000

22. A carpenter needed to cut 10 foot 2x4 into 5 or N equal sections. How many cuts will he have to make? Answer: 4, N-1

23. What are the next 3 numbers in the following set, assuming they follow the same pattern? 1,5,10, 16, 23. _, _, _. Answer: 31, 40, 50

24. If the fraction A/B where A and B are real numbers, then is the answer always smaller then A? (No and justify your answer. Note: One counter example will disprove a general conclusion.)

25. What is a possible conclusion to find the sum of x odd numbers from the following?

$1+3 = 4$
$1+3+5 = 9$
$1+3+5+7 = 16$
Ect:

Venn Diagram problem:

26. A survey reveals the following:

21 men drive Fords

25 drove General Motors cars

15 drove Chrysler products

4 were driving Ford and General Motors

3 were driving General Motors and Chrysler cars

5 were driving Ford and Chrysler cars

1 had all three company cars.

Questions. How many drove ONLY Fords cars, only General Motors cars, only Chryslers cars. Answer: 12,19, 8

27. What is a possible conclusion to find the sum of x odd numbers from the following?

$1+3 = 4$
$1+3+5 = 9$
$1+3+5+7 = 16$
$1+3+5+7+9 = ?$: Answer: The sum of the odds is a perfect square.

28. a. How many diagonals does a square have?
   b. How many diagonals does a regular pentagon have?
   c. How many diagonals does a regular hexagon have?
   d. How many diagonals does a regular heptagon have?

Do you see a pattern and predict how many diagonals a regular octagon has?

Answer: $D = N(N-3)/2$ where N is greater than 3.

29. Prediction or induction problem.

$5 \times 5 = 25$
$15 \times 15 = 225$
$25 \times 25 = 625$
$35 \times 35 = ?$ Predict and then ck your answer using calculator.
$45 \times 45 = ?$
Write a conclusion.

ACT Test problem:
30. Draw a 30-60 degree right triangle and indicate the altitude from the vertex of the right angle.
    a. Indicate the size of all 6 angles.
    b. Map the three similar triangles.
    c. If the hypotenuse is 10 units then calculate the lengths of all the segments.

Decision Making Review

As mentioned before the teaching of Mathematics has two major objectives:
1. Teaching the mathematics needed for whatever the professional background calls for. (This means being a student is a life time obligation.)
2. The skill of Decision Making for a better life. This was stated at the entrance to Plato's Academy as: **LET NO MAN ENTER HERE IGNORANT OF GEOMETRY**.

To meet the needs of modern day requirements this has been changed to:
LET NO STUDENT EXIT HERE IGNORANT OF GEOMETRY. The problem is that not many Geometry teachers teach their course with this objective.

31. Ask the student to list what they think everyday decisions are based on.
    A few items may be:
    > What they see. (TV)
    > What they read.
    > What somebody told them.(Is the person an authority?)
    > National Polls (What information should be stated? See below.)
    > Methods of logic, direct and indirect reasoning.
    > Bias
    > Tradition
    > Etc and keep a list for future reference.

    An example of the use and problems with polls was recently in Time magazine.
    January 28, 2013 issue.
    The following was stated with regard to the issue of Gun Control:
    > Would you favor a background check?

92% at the store 87% at gun show 75% private seller
Would you favorite the following for gun safety?
69% gun registration 58% ban on clips
56% ban on assault guns 52% ban on Ammo sells
Would you favor or oppose armed guards in schools?
54% favor 45% oppose
Do you own a gun?
49% Yes 49% no
Who is to blame for gun violence in the USA?
37% parents 37% pop culture 23% availability of guns

This is only a part of the survey, but what is very important in evaluating the results of any survey. The following was stated in very fine prints and hard to read.

The survey was listed as international, conducted by phone on Jan. 14& 15 (Sunday and Monday) of 814 adult Americans selected at random. Would you draw a general conclusion based on 814 adults without knowing:

1. Were they Gun owners?
2. Where they live?
3. Age
4. Profession or career.
5. Education
6. What is their favorite news show?

Discuss the responses to the above with your parents.

32. If you are a good teacher, then you like students (Assume this to be true)
    Which of the following are valid? (Remember: Statements are true or false and conclusions are valid or invalid.)

    a. John is a good teacher, then he likes students.
    b. Peter likes students, then he is a good teacher.
    c. Huck doesn't like students, the he is not a good teacher.
    d. Midge isn't a good teacher, then she doesn't like students.
         Answers: V, I, V. I (In order A-D)
    Use A→ B diagram to show the relationships for the converse, inverse and contrapositive.

33. From the following statement write two conclusion that are valid and two that are invalid.

    The State's Good Driver Association indicated that very few drivers were killed at speeds over 100 mph.

    Letters to the editor stated the many conclusion, some valid and some invalid.

    a. (your valid conlusion.)
    b. (your valid conclusion)
    c. (your invalid conclusion.)
    d. (your invalid conclusion)

34. What is inductive reasoning? The following will illustrate why you need to be careful when arriving at a conclusion by inductive reasoning. (Problem credited to Leo Moser)

    a. Draw a circle on your paper with at least a 2 inch diameter.
    b. Select 2 points, say A and B, on the circumference and draw segment AB.
       ? The circle is divided into how many regions. Conclusion: 2 points, then 2 regions.
    c. Pick another point, say C, on the circumference.
       ? How many regions now? Conclusion: 3 points and __ regions.
    d. Pick another point, say D, on the circumference.
       ? How many regions now? Predict: Conclusion: 4 points and __ regions by counting.
    e. Pick another point, say E, on the circumference.
       ? How many regions now? Predict? Conclusion: 5 points and __ regions by counting

       Was your prediction correct? What does this tell you with regard to inductive reasoning?

35. Indirect Reasoning How did A know he had a black hat on?
    Three students had a GPA of 4.0 and the class asked the teacher which one he thought is the most intelligent? The teacher set up the following test. He instructed each of the three to stand in a corner of the room. He then **blind folded** each of them and carefully explained he would put a **black or white** hat on each of them. When the blind folds are removed they are to put their

right arm above their head if they see a black hat and they were put their hand down when their know the color of their hat. He put a black hat on each and took off the blind folds. They each put their hand above their head. After a short while one student took he hand down and said I have a black hat on! How did he or she know? Suggestion: Have three black hats and act this out in class for complete understanding!

Index 1

Basic Definitions

(It is assumed some of the basic math. and geometric definitions are not necessary to define since you covered them in previous math classes.)

Point, line, ray, and plane are many times classified as undefined terms.

Point   •
Line    Ray

If you move or roll a point continually, both directions, a line is generated. In geometry, a line means a straight line.

If you move a point continually in one direction, a ray is generated. (Sometimes called a half line.)

Line segment

If you roll or move a point, then a line segment is generated.

Plane

A plane in boundless, so the above is called a plane segment.
A plane is generated by moving a line segment.

A ruler or straightedge, protractor, and calculator are the basic tools, even a computer could now be added.

These are just a few of the many definitions in a Geometry text.

Def. 1: A definition is valid, if the definition is true when reversed.
Def. 2: All valid conclusions are based on Undefined Terms, Defined Terms, Assumptions, and previously proven or accepted Conclusions.

Def. 3:   A triangle is the set of three non-collinear points and the line segments determined by the three points.

Def. 4:   Angle consists of two rays with the same beginning point.

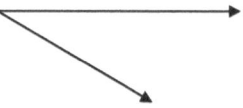

The method for labeling angles is by using three letters, one letter on each ray and one at the beginning point. The beginning point is called the vertex.

Example: ∠ADG where the middle letter is always at the vertex. Notice the symbol for angle (∠).

The drawing for ∠ADG is:

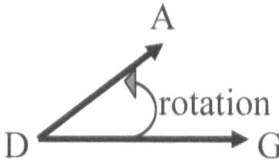

Related information:

An angle of 360 degrees (symbol for degree is °) in a complete rotation or circle.

A straight angle or half circle is 180°.

A right angle is 90°.

1° equals 60 minutes (symbol for minutes is '.)

The first set of minute parts.

1 minute equals 60 seconds (symbol for seconds is ".)

The second set of minute parts.

Def. 5:   An important conclusion in mathematics that is proven is called a theorem.

Def. 6:   Indirect proof is a method where you list all the possibilities and show that all but one is impossible, therefore the remaining possibility is the correct one.

Def. 7:   Two polygons figures are similar, if when mapped the corresponding angles are equal and the ratios of the corresponding parallel sides are equal.

The symbol that indicates the figures are similar is ~, like a lazy s.

Def. 8:   A polygon is a closed figure consisting of n points and the line segments determined by the n points, such that none of the line segments intersect except at their end points.

Def. 9:   A triangle is a three-sided polygon.

    a.   Triangles types are also acute, right, or obtuse.

    b.   A triangle is a rigid figure and its importance in construction

Def. 10: Congruency: If two polygons are similar and the ratio of the corresponding sides is 1, then the figures are congruent.

Note: Q.E.D (quod erat demonstrandum) is in one of the exercises It means: "That which was to be proved, has been."

Def. 11: An isosceles triangle is a triangle with two and only two equal sides.

Def. 12: A rectangle is a parallelogram with right angles.

Def. 13: A parallelogram is a four-sided polygon with the opposite sides parallel.

Def. 14: A rectangle is a parallelogram with right angles.

Def. 15: The AREA of a geometric figure is the number of square units contained in the interior of a figure.

Def. 16: In a right triangle the side opposite the right angle is the hypotenuse.

Def. 17: In a triangle the altitude is the line segment from the vertex of the angle perpendicular to the opposite side. (The side may have to be extended.)

Def. 18: In a triangle, a median is defined as the line segment from the vertex of the angle to the midpoint of the opposite side.

Def. 19: The four forms of "If A, then B." are:

Theorem	If A, then B.
Converse	If B then A.
Inverse	If not A, then not B.
Contrapositive	If not B, then not A.

Def. 20: A circle is defined as a plane figure consisting of set of all points on the plane that is a given distance from a given point. (The given point is labeled the center and the given distance is called the radius.)

Def. 21: Circle terminology

    a.   The given point in the above definition is called the center.

    b.   The line segment from the given point (center) to a point on the circle is called the radius.

    c.   The distance around the circle is the circumference.

d. The line segment from one point on a circle to another point on the circle is called a chord.

e. If a chord contains the center point, it is called a diameter.

f. A line that intersects a circle in only one point is called a tangent.

g. A secant is a line that intersects a circle in two points.

h. A portion (segment) of a circle is called an arc.

Def. 22: A circle's central angle is an angle with its vertex at the center of the circle.

Def. 23: The measure of the arc a central angle intercepts is the same measure as the central angle.

Def. 24: An inscribed angle is an angle where the vertex is on the circumference of a circle and the rays or sides are chords.

Concept: 3-D figures and layout view

Def. 25: Volume is the number of cubic units a 3-D figure contains.

Def. 26: A sphere is the set of all points equidistant from a given point, called the center. (See Def. 20)

Def. 27: The SINE (Sin) of an acute angle in a right triangle is the ratio of the length of the side opposite the angle divided by the length of the hypotenuse.

Def. 28: The COSINE (Cos) of an acute angle in a right triangle is the ratio of the length of the adjacent side divided by the length of the hypotenuse.

Def. 29: The Tangent of an acute angle in a right triangle is the ratio of the length of the side opposite the angle divided by the length of the side adjacent to the angle.

Def. 30: Terms and forms of an implication and their validity:

	Name	Symbolic	
Theorem or statement:			If True and valid

A → B

Read: If A, then B.

Converse: B → A   then truth and validity in doubt.

Read: If B, then A.

Inverse: ~A → ~B   then truth and validity in doubt.

Read: If not A, then not B. The negation symbol, read not, is ~.

Contrapositive: ~B → ~A   then true and valid

Read: If not B, then not A.

Def: 31: The MEAN or arithmetic average for a set of numbers is the sum of the numbers divided by n, the number of numbers in the set. Formula: Mean = (sum of n scores)/n

Def: 32: The MODE for a set of data is the most popular or most frequently occurring element or score in the set.

Def: 33: The MEDIAN for a set of data is the middle element or score when the elements or scores are arranged in order from lowest to highest.

Note: Q.E.D. (quod erat demonstrandum) is a term often used in justifying a view point. It means: "That which was to be proved, has been."

Comment: many students were taught the Order Of Operations, which is to simplify a set of arithmetic operations you do the multiplication first and then the additions. You may be wondering where the operations of division and subtraction are? The answer is the division is really multiplication (dividing by 3 is really multiplying by 1/3, and subtracting 3 is really adding-3.) To illustrate this place a pile of coins (Pennies, nickels, dimes and quarters on the table and then count them to determine the value.) You will recognize that you automatically use the definition of the order of operations

Index 2

Postulates

Listed below are the postulates for a logical Geometric system that are only used in this review.

Post. 1: A line has an infinite set of points.

Post. 2: Two points will determine one and only one straight line.

Post. 3: There is a one to one correspondence between the points on a line and the real number line.

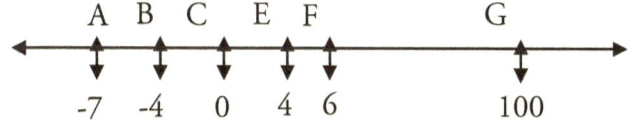

(Postulate 3 enables you to use algebra to help solve geometric problems.)

Post. 4: The shortest distance between two points is the measure of the straight line segment. (This is not always true as you saw in the taxicab exercise in Session 1, and that is why it is a postulate)

Post. 5: The shortest distance between a point and a line (on a plane) is the perpendicular line segment.

Post. 6: Three non-collinear points will determine a plane.

Post. 7: If two parallel lines are crossed by another line (called a transversal), then the alternate interior angles are equal.

Post. 8: Through a point not on a given line, there is only one line through the point that is parallel to the given line.

Post. 9:   The area of a rectangle is equal to the length times the width and the answer is in square units. (The basic unit of measure for base and height must be the same.) Formula: A = bh sq. units.

Post. 10: The formula for the circumference of a circle is C = 2πr or πd. (This can be proved, but not in most H. S Gemetry texts.)

Post. 11: The formula for the area of a circle is A = πr². (This can be proved.)

Post: 12: The volume of a rectangular solid is the length times the width times the height, or V = LWH and the answer is in cubic units. (L, W, and H must be in the same units of measure.)

Index 3

Essential Geometry
Theorems

(All of these theorems can be proved, but most students and users only remember the formulas and how to use them.)

Th. 1:   In a triangle the sum of two sides is greater than the third side, and the third side is greater than the difference of the two sides

Th. 2:   If two lines intersect, then the opposite angles are equal.

Th. 3:   The sum of the angles in a plane triangle is 180 degrees.

Th. 3:   The exterior angle of a triangle is equal to the sum of the two non-adjacent interior angles.

Th. 5:   I f two lines are crossed by another line (transversal) so that the alternate interior angles are equal, then the lines are parallel.

Th. 6:   If two angles of one triangle are equal in measure to two angles of another triangle, then the triangles are similar and all the conditions in the general similarity definition are valid.

Th. 7:   If the corresponding sides are in equal ratio, then the triangles are similar and all the conditions in the general similarity definition are valid.

Th. 8:   If two triangles have two corresponding sides proportional and the included angles equal, then the triangles are similar and all the conditions in the general similarity definition are true.

Th. 9:   In an isosceles triangle, the angles opposite the equal sides are equal in measure. (Usually called the base angles.)

Th. 10:   The angles in an equilateral triangle are equal in measure.

Th. 11:   In a scalene triangle, there is a correspondence between the length of the sides and the angles opposite the sides. The larger the side, the larger the opposite angle.

Th. 12:   The diagonals in a parallelogram bisect each other.

Th. 13:   In a parallelogram the opposite angles are equal and the opposite sides are equal in measure.

Th. 14:   The diagonals of a rectangle are equal in measure.

Th. 15:   The area of a triangle is ½ the measure of the base times the measure of the height or ½ base times the altitude. Formula: $A_\triangle$ = (½)b)h sq. units.

Th. 16:   The area of a parallelogram is equal to base time the height, or base times the altitude. Formula: A ▱ = bh sq. units

Th. 17:   The area of a trapezoid is the sum of the two bases times ½ the the measure of the altitude.
           Formula is A ◺ = (½)h(b_1 + b_2) sq. units

Th. 19:   Any point on the perpendicular bisector of a line segment is equal distant from the end points of the segment.

Th. 20:   In a right triangle with sides a, b, and c is the hypotenuse, then $a^2 + b^2 = c^2$. **VIT (Very Important Theorem)**

Th. 21:   In a 30-60 degree right triangle the side opposite the 30 degree angle is ½ the hypotenuse.

Th. 22:   In a 30-60 degree right triangle the side opposite the 60 degree angle is half the hypotenuse times the $\sqrt{3}$ or (h/2)$\sqrt{3}$.

Th. 23:   The three altitudes of a triangle intersect at a point (are concurrent), but the sides may have to be extended.

Th. 24: The three medians of a triangle are concurrent at a point 2/3 the length of each median.

Th. 25: The angle bisectors of a triangle are concurrent at a point equal distant from the sides.

Th. 26: If a point is equidistant from the sides of an angle, then the point is on the bisector of the angle.

Th. 27: The segment from the center of a circle to the point of tangency is perpendicular to the tangent.

Th. 28: The tangent from a point outside the circle squared, plus the radius squared, equals the square of the distance from the point to the center of the circle.

Th. 29: The perpendicular bisectors of the chords of a circle intersect at the center of the circle.

Th. 30: An inscribed angle in a circle is equal to ½ the degree measure of its central angle or ½ the degree measure of the intercepted arc. (See Def. 12.2)

Th. 31: In a plane figure if the dimensions are doubled, then the area is increased 4 times.

Th. 32: If the diagonals of a parallelogram are equal, then the parallelogram is a rectangle.

Th. 33: If a line is perpendicular to two lines in a plane at the point of intersection, then it is perpendicular to the plane.

Th. 34: The volume of prism or cylinder is the product of the area of the base times the height, $V = A(h)$ cu. units. (Height or altitude is the measurement of the perpendicular distance. The measurements must be in the same units.)

Th. 35: The volume of a pyramid is 1/3 the area of the base times the height. The formula is VPyramid = Bh/3 cu. units (B and h in the same units)

Th. 36: The volume of a cone is 1/3 the area of the base times the height or altitude. The formula is $A_L = \pi r^2 h/3$. cu. units (r and h in the same units. Only circular bases were considered.)

Th. 37: The formula for the lateral area of a cone is A = (1/2) Ch sq. units, where C is the Circumference of the base and h is the slant height.

Th. 38: The formula for the volume of a sphere is $V = (4/3)\pi r^3$ cu. units, where r is the radius.

Th. 39: The surface area of a sphere is $4\pi r^2$ sq. units, where r is the radius.

Th. 40: The ratio of the areas of two spheres is the square of the ratio of their radii. $a/A = (r/R)^2$

Th. 41: The ratio of the volumes of two spheres is the cube of the ratio of their radii. $v/V = (r/R)^3$

Th. 42: One formula for Pi or π is the limit of n(sin 180/n) as n gets very large. Where n is the number of sides in the inscribed regular polygon.

The four forms of an implication and how they are related. These were "justified" by set diagrams.

Statement	Conclusion
Theorem (A→B)	If True, and valid then
Converse (B→A)	Not necessarily true or valid
Inverse (~A→~B)	Not necessarily true or valid
Contrapositive (~B→~A)	True and valid

Logic statement: Statements are true or false, but conclusions are valid or invalid!

Index 4

Conversion Information

Linear

English		Metric or International (SI)
1 inch	=	2.54 centimeters (cm.)
1 foot = 12 inches	=	30.48 cm.
3 feet = 1 yard	=	.9144 m. or 91.44 cm.
3.2808 ft or 39.37 inches = 1 meter	=	10 decimeters (dm.) = 100 cm.
5280 ft. = 1760 yds = 1 mile	=	1.609 kilometers = 1609 meters
Radius of earth = 3963 miles	=	6377 km.

Area (Approximations)

1 square inch	=	6.45 sq. cm.
1 square foot = 144 sq. in.	=	929.03 sq. cm.
1 square yard = 9 sq. ft.	=	5625 sq. cm. or .863 sq. meters
10.76 sq. ft. = 1.196 sq. yds	=	1 sq. meter = 10,000 sq. cm.
2.47 acres	=	1 hectare = 10,000 sq. meters
1 acre = 43,560 sq ft or 4,840 sq. yds.	=	4,177 sq. meters = .4177 hectares

Volume (Approximations)

1 cubic inch	=	16.39 cu. cm.
1 cu. ft. = 1728 cu. in.	=	.028 cu. meters
1 cu. yd = 27 cu. ft.	=	.765 cu. meters
35.31 cu. ft = 1.31 cu. yds.	=	1 cu. meter
1 quart	=	.9463 liters
1 gallon = 4 qts.	=	3.785 liters
1.0568 U.S. gallons	=	4 liters or 1 Canadian gallon

1 cubic foot of water is 62.43 lbs. or 28.31 kilograms
1 gal of water weighs 8.35 lbs.

Angle Measurement

1 minute = 60 seconds
1 degree = 60 minutes   =   .01745 radians
57.2957 degrees(rounded)   =   1 radian
180 degrees   =   3.1416 (Π) radians (rounded)
One revolution equals 360 degrees or 2Π radians

Index 5

Suggestions for further reading

** Indicated selections highly recommended for students. You will be surprised.

Abbott, Edwin A.
FLATLAND A ROMANCE OF MANY DIMENSIONS
Princeton University Press

Banks, Robert B.
SLICING PIZZAS,RACING TURTLES,AND FURTHER ADVANTURES IN APPLIED MATHEMATICS
Princeton University Press

Beckmann, P.
HISTORY OF PI
Golden Press

Bell, E. T.
MEN OF MATHEMATICS
Simon & Schuster

Byrkit, D.
"TAXICAB GEOMETRY." **
MATHEMATICS TEACHER, May 1971, Pages 418-422

Cajori, Florian
HISTORY OF ELEMENTARY MATHEMATICS
The Macmillan Company

Davis, J.J.
BIBLICAL NUMEROLOGY++
Baker Book House (PI value is stated in the Bible, erroneously, I Kings 7:23)

Davis, P. and Hersh, R.
THE MATHEMATICAL EXPERIENCE
Houghton Mifflin

DESCARTES' DREAM
Harcourt Brace Javanovich

Devlin, K
Mathematics-the new golden age
Columbia University Press

Dudley, Underwood
NUNEROLOGY or, What Pythagoras Wrought
Mathematical Association of America

MATHEMATICAL CRANKS
Mathematical Association of America

Fadiman, Clifton
THE MATHEMATICAL MAGPIE
(Mobius Strip-"Paul Bunyan vs. The Conveyor Belt)"
Simon and Schuster

Fawcett, Harold
NATURE OF PROOF
13th Yearbook of NCTM

Florman, S. C.
ENGINEERING AND THE LIBERAL ARTS
McGraw-Hill Company
(A guide to History, Literature, Philosophy, Art, Science, and Music)

Gardner, M
MATHEMATICAL CARNIVAL**
Alfred A. Knopf

MATHEMATICAL CIRCUS**
Vintage Books
Division of Random House

Gazale, Midhat
NUMBER: From Ahmes to Cantor
Princeton University Press

Gordon, Sheldon and Florence, Editors
STATISTICS FOR THE TWENTY-FIRST CENTURY
Mathematical Association of America, 1992

Huff, Darrell
HOW TO LIE WITH STATISTICS**
Norton Co.

Kenny
"Hemholtz And The Nature Of Geometric Axioms"**
Mathematics Teacher, Vol. 50, Feb. 1957

Klein H. A.
THE WORLD OF MEASUREMENTS
Simon and Schuster

Kline, M.
MATHEMATICAL THOUGHT FORM ANCIENT TO MODERN TIMES
Oxford University Press

Lieber, L.
MITS, WITS, AND LOGIC**
Institute Press, New York, 1954

THE EDUCATION OF T. C. MITS**
W. W. Norton & Co., 1954

Loomis,E.
THE PYTHAGOREAN PROPOSITION.
NCTM publication
Comment: (Which former President of the U.S. is credited with a proof?)

Nolan, Deborah, Editor
WOMEN IN MATHEMTICS: SCALING THE HEIGHTS
Mathematical Association of America

Northrop, E. P.
RIDDLES IN MATHEMATICS** (A Book of Paradoxes)
D. Van Nostrand Company

Packel, Edward
THE MATHAMATICS OF GAMES AND GAMBLING**
Mathematical Association of America

Paulos, J.
I THINK, THEREFORE I LAUGH**
Vintage Books
Division of Random House

Peterson, I.
THE MATHEMATICAL TOURIST**
W. H. Freeman and Company

Poe, Edgar Allen
THE GOLD BUG** (A Mystery involving mathematical reasoning.)

Polya, G.
MATHEMATICAL DISCOVERY: Vol. 2 (Chapter 14: The art of teaching mathematics.)
John Wiley & Sons

Postman, N.
TECHNOPOLY**
Alfred A. Knopf

Reid, Constance
A LONG WAY FROM EUCLID**
Thomas Y. Crowell Co.

Reeve, W. D.
THE TEACHING OF GEOMETRY
5th Yearbook NCTM

Stevenson, R. L.
TREASURE ISLAND (chapter 31)**
(Locus problem-location of the treasure.)

Weber, R.
A RANDOM WALK IN SCIENCE**
Crane, Russak & Co. Inc.
    "Life on Earth.(by a Martian")
(Fascinating little story (p. 124) with a surprise ending.

Video or film (May be available at the library or they can order it.)
DONALD DUCK IN MATHMAGIC LAND
Disney

An interesting critical thinking test. (Your teacher may be interested in this test.) Critical Thinking Test, Level X

R. Ennis and J. Millman
    (Very interesting and a different type of test based on a space travel theme.
    I have given this to several hundred high school and college students on a pre/post test situation and to my surprise the average group gained the most.)

Available at:
    Foundation for Critical Thinking 1-800-833-3645 or 1-800-458-4849
    www.criticalthinking.org

Web sites

www.//history.mcs.st www.MAA.org
www.archives.math.utk.edu/societies.html
www.nsf.gov/ www.AMS.org/ www.forum.swarthmore.edu/ncsm
http://Turnbull.mcs.st www.history.mcs.st-andrews

Also use computer search for "Math History" or "Math Archives"

Index 6

Quotes

Pertaining to mathematics
and decision making

Mathematics is the gate and the key to all sciences. He who is ignorant of it cannot know the things of this world.

Roger Bacon

Young people who have acquired the ability to analyze problems, gather information, put the pieces together to form tentative solutions will always be in demand.

J. G. Maisonrouge
Board Chairman
IBM World Trade Corp

That they (all citizens) might excel in public discussions on philosophic or scientific questions, they must be educated (rhetoric, philosophy, mathematics, and astronomy).

The Athenian Sophist School Curriculum(480 B.C.E.)
F. Cajorie

Consciously mathematics has been a human activity for thousands of years. to some small extent, everybody is a mathematician and does mathematics.

Phillip Davis & Rueben Hersh
THE MATHEMATICAL EXPERIENCE

GOD gave us the integers (whole numbers) and all the rest is the work of man.

L. Kronecker

Mathematics is not a spectator sport!

Anonymous

Neglect of mathematics works injury to all knowledge.

Roger Bacon

Number rules the universe.

The Pythagoreans

Mathematics-the unshaken Foundation of Sciences, and the plentiful Fountain of Advantage to human affairs.

Issac Barrow

Understanding evolves from work, appreciation from applications.

Unknown

Geometry with too much rigor only produces rigor mortise.

F. Allen

The theory of probability entered mathematics through gambling.

P. Davis & R. Hersh
THE MATHEMATICAL EXPERIENCE

In truth, all of life in one way or another is concerned with the study of probability.

H. Gross & F. Miller
MATHEMATICS-A Chronicle of Human Endeavor

In short, the house plays the percentages, while the player relies on luck . . .

H. Gross & F. Miller
MATHEMATICS-A Chronicle of Human Endeavor

All numbers in the form of 4n+1 are the sum of 2 squares.

Fermat

A mathematician, like everyone else, lives in the real world. But the objects with which he works do not. They live in that other place-the mathematical world. Something else lives here also. It is called TRUTH.

Jerry P. King
THE ART OF MATHEMATICS

You cannot fake. In mathematics, no one can be fooled. You can either prove . . . or you cannot.

Jerry P. King
THE ART OF MATHEMATICS

Many of the laws of the sciences are stated in the language of variation.

Unknown

Mathematics is like a mighty tree with number (counting numbers) for its roots. Arithmetic grows on numbers, algebra on arithmetic, geometry on arithmetic and algebra, analytic geometry on arithmetic, algebra, and geometry. Calculus builds on all four. It is a tree thatgrows in time, fertilized by the minds of mathematicians and the applied needs of society.

Unknown

Attributing teaching and learning failure to something called "math anxiety" serves no purpose except to provide a built-in excuse for inadequate performance on both sides.

Jerry P. King
THE ART OF MATHEMATICS

Statistical thinking will one day be as necessary for efficient citizenship as the ability to read and write.

H. G. Wells

Pythagoras, the teacher, paid his student three oboli (a coin) for each lesson he attended and noticed that as the weeks passed the boy's initial reluctance to learn was transformed into enthusiasm for knowledge. To test his pupil Pythagoras pretended that he could no longer afford to pay the student and that the lessons would have to stop, at which point the boy offered to pay for his education . . .

Simon Sing
FERMAT'S ENIGMA

TO MEASURE IS TO KNOW

Johann Kepler

Hipparchus of Nicaea, (180-125 B.C.E.) compiled the first trigonometric table.

Boyer, C.B.
A HISTORY OF MATHEMATICS

The advance and the perfecting of mathematics are closely joined to the prosperity of a nation.

Napoleon

The heart of the mathematical experience is, of course, mathematics itself.

Davis, P and Hersh, R
THE MATHEMATICAL EXPERIENCE

Mathematics through the power of computers pervades almost every aspect of our lives . . .

David L. Goines

To Think is to Know

Unknown

Let no man ignorant of Geometry enter here.

Plato

Let no person ignorant of Geometry exit here.

J. Elander

Students of mathematics . . . the first time something new is studied seem they hopelessly confused . . . Then, upon returning (to the concept) after a rest, . . . everything falls in place.

E. T. Bell
MEN OF MATHEMATICS

Descartes . . . the essence of plane analytic geometry lies in the matching of ordered pairs of real numbers with points on a plane.

Edna E. Kramer
THE NATURE AND GROWTH OF MODERN MATHEMATIC

Thinkers recognize when two variables are related, but it is mathematics that connect them numerically.

Unknown

There cannot be a language (mathematics) more universal . . . and more worthy to express the invariable relations of the natural things.

Joseph Fourier

Analytic Geometry . . . constitutes the greatest single step ever made in the progress of the exact sciences.

John Stuart Mill

I THINK THEREFORE I AM.

Rene Descartes

The Great Architect of the Universe now begins to appear as a pure mathematician.

J.H.Jeans
The Mysterious Universe

The definition of a good mathematical problem is the mathematics it generates rather than the problem itself.

Andrew Wiles

We learn the new in the light of the old.

Anonymous

The important thing is to not stop questioning . . .
                                            Albert Einstein

Relationships between different subjects (even branches of
mathematics) are creatively important in mathematics.
                                            Simon Singh
                                        FERMAT'S ENIGMA

Mathematics consists of islands of knowledge in a sea of
ignorance.
                                            Simon Singh
                                        FERMAT'S ENIGMA

Statistics makes possible new perceptions and realities by
making visible large-scale patterns.
                                            Neil Postman
                            Technopoly-The Surrender of Culture
                                            to Technology

Statements are true or false. Conclusions are valid or
invalid
                                            Unknown

Just as statistics has spawned a huge testing industry, it has
done the same for the polling of "public opinion."
                                            Neil Postman
                            Technopoly-The Surrender of Culture
                                            to Technology

It is not how much you cover, but how much you uncover.
                                            H. Fawcett

The connection between the improvement of human
conditions and the happiness of the human race is Science.
(The Queen of the sciences is MATHEMATICS.)
                                            Neil Postman
                            Technopoly-The Surrender of Culture
                                            to Technology

The proof of the pie is in the eating.

Unknown

There is no royal road to Mathematics.

Menaechmus (to Alexander the Great)

Mathematics is the science of making necessary conclusions.

B. Peirce

It is easier to square the circle then get around a mathematician (This is one of the three famous problems of Geometry.)

A. DE Morgan

Mathematics is about anything as long as it is a subject that exhibits the pattern of assumption-deduction-conclusion.

P. Davis and R. Hersh
The Mathematical Experience